SPORT COMPACT
CUSTOMIZING

Haynes Publishing Group
Sparkford Nr Yeovil
Somerset BA22 7JJ
England

Haynes North America, Inc
861 Lawrence Drive
Newbury Park
California 91320 USA

Acknowledgements

We are grateful for the help and cooperation of APC (American Products Company) who supplied many of the custom car photographs used throughout this book, including the front cover and to Charlie Kim of Dynamic Autosports Inc., and Miguel Ortiz of M.O.B. works who provided valuable insight into the world of Sport Compact customizing.

© Haynes North America, Inc. 2003
With permission from J.H. Haynes & Co. Ltd.

All rights reserved. No part of this book may be reproduced or transmitted in any form of by any means, electronic or mechanical, including photocopying, recording or by any information storage or retrieval system, without permission in writing from the copyright holder.

ISBN 1 56392 505 2
Library of Congress Control Number 2003107915

Printed by J H Haynes & Co Ltd,
Sparkford, Yeovil, Somerset BA22 7JJ, England.

While every attempt is made to ensure that the information in this manual is correct, no liability can be accepted by the authors or publishers for loss, damage, or injury caused by any errors in, or omissions from, the information given

03-184

Be **careful** and know the **law**!

1 Advice on safety procedures and precautions is contained throughout this manual, and more specifically within the Safety section towards the back of this book. You are strongly recommended to note these comments, and to pay close attention to any instructions that may be given by the parts supplier.

2 Haynes recommends that vehicle modification should only be undertaken by individuals with experience of vehicle mechanics; if you are unsure as to how to go about the modification, advice should be sought from a competent and experienced individual. Any queries regarding modification should be addressed to the product manufacturer concerned, and not to Haynes, nor the vehicle manufacturer.

3 The instructions in this manual are followed at the risk of the reader who remains fully and solely responsible for the safety, roadworthiness and legality of his/her vehicle. Thus Haynes is giving only non-specific advice in this respect.

4 When modifying a car it is important to bear in mind the legal responsibilities placed on the owners, drivers and modifiers of cars. If you or others modify the car you drive, you and they can be held legally liable for damages or injuries that may occur as a result of the modifications.

5 The safety of any alteration and its compliance with construction and use regulations should be checked before a modified vehicle is sold as it may be an offense to sell a vehicle which is not roadworthy.

6 Any advice provided is correct to the best of our knowledge at the time of publication, but the reader should pay particular attention to any changes of specification to the vehicles, or parts, which can occur without notice.

7 Alterations to a vehicle should be disclosed to insurers and licensing authorities, and legal advice taken from the police, vehicle testing centers, or appropriate regulatory bodies.

8 Various makes of vehicle are shown being modified. Some of the procedures shown will vary from make to make; not all procedures are applicable to all makes. Readers should not assume that the vehicle manufacturers have given their approval to the modifications.

9 Neither Haynes nor the manufacturers give any warranty as to the safety of a vehicle after alterations, such as those contained in this book, have been made. Haynes will not accept liability for any economic loss, damage to property or death and personal injury other than in respect to injury or death resulting directly from Haynes' negligence.

Contents

01

From here to there 6

02

The bad with the good 8

03

Security 10
Installing a Basic LED 14
Wiring Basics 15
Installing an Alarm 16

06

07

08

Mobile Entertainment 96
In-dash Receivers/Players 98
Speakers 100
Amplifiers 107
Subwoofers 110
Video 114

Suspension 118
Struts and Coil Springs 124
Rear Coil Spring
 Suspension 128
Coilovers 131
Air Suspension 132
Nasty Side Effects 133
Strut Braces 134
Stabilizer Bars 136

Wheels and Tires 140
Gallery of Wheels 144

Interiors 22
Sport Steering Wheels 24
Gearshift Knobs and
 Boots 30
Custom Pedals 36
Floor Mats 39
Neon Lighting 40
Interior Trim 44

Gauge Face Upgrade 48
Installing a Four-point
 Harness 55
Bucket Seats 58
Starter Button 60
Installing Gauges 63
Window Tinting 66

Body and Exterior 68
De-Badging 70
Aftermarket Mirrors 73
Color-matching Exterior
 Accessories 76
Custom Taillight Lenses 78

Install a Custom Grille 81
High-power headlights 84
Rear Wings and Spoilers 86
Body Kits 88
Neon 93
Custom Painting 94

Brakes 146
Upgraded Discs and
 Pads 148
Bigger Brakes 152
Painting Calipers 156
Painting Drums 157

Engine Performance 158
Exhaust Systems 160
Turbochargers 163
Superchargers 166
Nitrous Oxide 169
Induction Systems 172
Computers and Chips 174
Valvetrain Modifications 176
Ignition Upgrades 178

Safety First 180

Source List 182

From **here . . .**

01

From here . . . to there

Go to a sport-compact car show. You'll see the racer with grease under his fingernails, bent over his hood, adjusting his turbo wastegate. You'll see the themed car with matching colors (even under the hood) with a mirror-like paint job reflecting its neon lights. You'll see the rolling sound cannons, with beautifully integrated audio systems, blasting at Richter-scale levels. But you'll also see the daily-driven street cars with custom wheels, body kits and graphics that make them stand out from the other snore-mobiles in the neighborhood. This is truly diversity. And there is no "right" way to customize a car.

. . . to **there**

It seems our cars reflect our personalities, and that's the way it should be. If you're the "all-business" type, you'll probably not care much about your paint job and put the money you save into your engine. For others, looks and appearance are more important, but few of us have the money to "do it all." It's best to figure out what's most important to you and do that first. Look at the number of "unfinished" cars at any show; it illustrates how trying to do it all at once can backfire, especially when you run out of time and money. Most experienced car builders will plan "stages" of modification, each of which can be completed in relatively short periods of time. So, for example, you can get your paint job and graphics, then take some time to save up for your 12-second engine. In the meantime, you can enjoy driving a great-looking car. A die-hard racer might want to do it in reverse, but the principle is the same: it's best not to take on too much at once, unless of course you're the type who wants it all and wants it now - which isn't necessarily a bad thing either.

The Bad
with the Good

Customizing your car can get complicated in a hurry. What the manufacturers of high-performance and custom parts don't tell you is that there's usually a price to be paid for each "upgrade" you decide to make. For example, if you install big wheels and low-profile tires, be ready for a slightly rougher ride and be sure to stay away from potholes and curbs, since these wheels damage easily. If you install an intake tube, you're going to start hearing the air flowing into your engine (is that a good thing or a bad thing?). If it's a cold-air intake, you're going to have to watch out for deep puddles, or you might wind up with water in your engine (definitely a bad thing). If you lower your car, you're more likely to damage your suspension or destroy a tire from "fender rub." And speed bumps will become your worst enemy. Well, you get the idea. Your sport compact car was extremely well designed as it came from the factory. It was designed to provide long life, a comfortable ride and excellent fuel economy. These are all attributes we'd like to keep, if possible. So talk to as many people as you can who've actually done the modifications you're planning. Chances are they'll tell you some of the drawbacks that didn't show up in the magazine ad. And, on the other hand, they may let you know about some pleasant surprises they discovered after adding some custom pieces. Sometimes, you just don't know for sure, which is why you need to go in prepared to accept a little of the bad with the good!

Here are some common issues you may run into:

Component	For	Against
Performance computer chip	Increased power, realize benefits of other engine modifications.	May affect driveability and ability to pass emissions test.
Cat-back exhaust system	Slight gain in power, especially with other modifications. Louder (see also: Against).	Possible loss of ground clearance; Louder (see also: For).
Exhaust header	Slight gain in power, especially when combined with other modifications.	More exhaust noise; less ground clearance (with some designs); more possibility of exhaust leaks.
Power adders (nitrous oxide, turbocharging and supercharging)	Large power increase without tearing deeply into engine.	Greater chance of engine damage or short engine life if not properly set-up. Can cause you to fail an emissions inspection.
Performance camshaft	Significant power increase, especially when combined with other flow-improving modifications.	On bigger cams, rough idle, loss of low-rpm power; loss of engine vacuum (so power brakes may work poorly); Can cause you to fail an emissions inspection.
Nitrous oxide	Big power boost.	Power can come at the expense of engine components not up to the task.
Custom paint	Ultimate statement.	Expensive. At re-sale, will need to find someone else with exactly the same taste as you.
Body kit	Turns any common run-of-the-mill car into something unique and personal.	Can look cheap if paint and fit are not perfect.
Window tint	Help stop sun-fade of interior. More difficult for thief to see what goodies you have. Gives clean exterior look.	May or may not be 100% legal in your area.
Custom bucket seats	New look to the interior. Perfect final touch to other, more subtle, treatments. Can be more comfortable and provide better support over stock.	May not be compatible with standard seatbelt systems. Installation can be difficult if not designed specifically for your car.

03 Security

Avoiding trouble

Those shiny wheels and flashy paint are like a billboard to car thieves and bandits looking for expensive sound system components to sell on the black market. And you've got to be careful when and where you choose to show off your car's mobile entertainment, and to whom. Be especially discreet the nearer you get to home - turn your system down before you get near home, for instance, or you might draw unwelcome attention to where that car with the loud stereo's parked at night.

If you're going out, think about where you're parking - somewhere well-lit and reasonably well-populated is the best bet.

If you're lucky enough to have a garage, use it. And always use all the security you have, whenever you leave the car, even if it's a tedious chore to put on that steering wheel lock. Just do it.

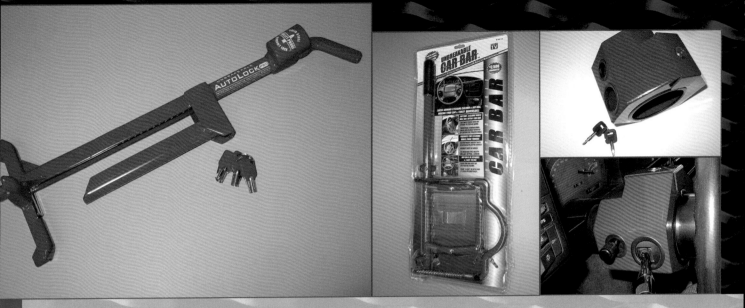

Anti-theft devices

Other types of anti-theft devices are available as a less expensive alternative to alarms.

An automobile equipped with a steering wheel lock or a removable steering wheel could make your car a less likely target for a thief. Also available are locking covers for the steering column which can help prevent a thief from being able to access the ignition lock, and devices that prevent the brake or clutch pedal from being depressed. Whatever your choice may be, now every time you park, at least you can relax a little. Remember, though, there's no guarantee that installing an alarm or security device will make any difference to a determined thief or mindless vandal.

If your vehicle is equipped with an alarm system or an anti-theft device, you may be eligible for discounted insurance premiums. Certain companies offer a higher percentage discount for vehicles that have more sophisticated alarm system packages. Each insurance company will have their own guidelines and insurance discounts. Contact your insurance representative for all the specific details.

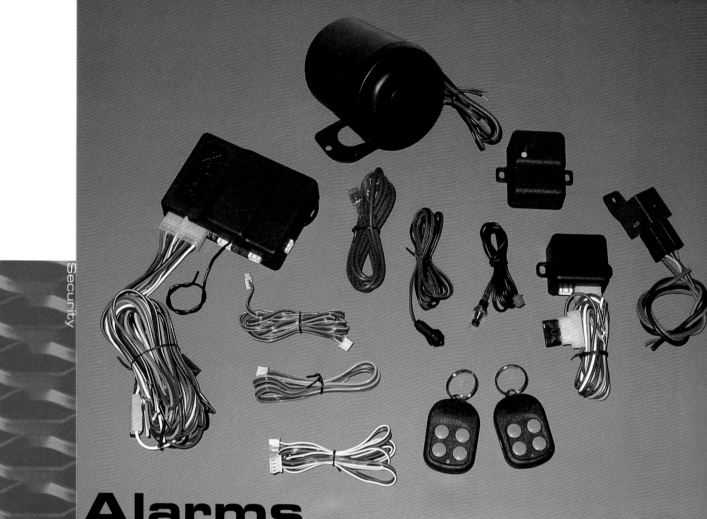

Alarms

Alarm systems are available in many different packages depending on cost and complexity. Here are just a few examples of types of alarm system sensors designed for different types of protection.

A more sophisticated alarm will feature shock sensing (which could be set off by a thief attempting to steal your tires); this type of sensor monitors the impact/vibration level.

Ultrasonic sensors monitor an enclosed space inside of a vehicle (such as the passenger compartment) with ultrasonic sound waves. If the sensor detects a change in the sound waves the alarm will sound.

Field Disturbance Sensors protect an area with an energy shield. Similar to ultrasonic sensors, FDS sensors monitor the space inside of a vehicle. They are often used to protect the passenger compartment of a convertible with the top down.

Pin switches typically monitor the doors, hoods or trunks by completing the circuit and activating the alarm when one of these has been opened.

Many suppliers have designed security systems not only to protect a vehicle; some are designed to give added convenience with packages that allow you to add onto the alarm network, such as keyless entry, power window control or remote starting capability. When purchasing an alarm, be sure to take the time and study each kit and their advantages and shortcomings.

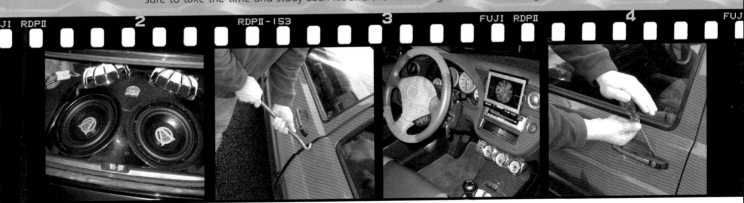

Installing a basic LED

All you need for this is a permanent live feed, a ground, a switch if you want to be able to turn it on/off, and the flashing LED itself (very cheap, from any car accessory shop).

An LED draws very little current, so you'll be quite safe tapping into almost any live feed you want. If you've wired in your stereo system, take a live feed from the permanent (radio memory supply) wire at the back of your head unit, or have a delve into the back of the fusebox with your test light. A ground can easily be tapped again from your head unit, or you can make one almost anywhere on the metal body of the car, by drilling a small hole, fitting a self-tapping screw, then wrapping the bared end of wire around and tightening it.

The best and easiest place to mount an LED is into one of the many blank switches the manufacturers seem to love installing. The blank switch is easily pried out, and a hole can then be drilled to take the LED (which usually comes in a separate little holder). Feed the LED wiring down behind the dashboard to where you've tapped your live and ground, taking care not to trap it anywhere, nor to accidentally wrap it around any moving parts.

Connect your live to the LED red wire, then rig your ground to one side of the switch, and connect the LED black wire to the other switch terminal. You should now have a switchable LED. Tidy up the wiring, and mount the switch somewhere discreet, but where you can still get at it. Switch on when you leave the car, and it looks as if you've got some sort of alarm - better than nothing.

Wiring basics

If you were thinking of taking an alarm live supply direct from the battery - don't. It's better to trace the red lead down to the starter motor, and tap in there. If a thief manages to get past your hood switch, the first thought will be to cut every additional live feed at the battery.

With your wires identified, how to tap into them? The three best options are:

a Soldering - avoids cutting through your chosen wire - strip away a short section of insulation, wrap your new wire around the bared section, then apply solder to secure it. If you're a bit new to soldering, practice on a few offcuts of wire first.

b Bullet connectors - cut and strip the end of your chosen wire, wrap your new one to it, push both into one half of the bullet. Connect the other end of your victim wire to the other bullet, and connect together. Always use the "female" half on any live feed - it'll be safer if you disconnect it than a male bullet, which could touch bare metal and send your car up in smoke.

c Block connectors - easy to use. Just remember that the wires can come adrift if the screws aren't really tight, and don't get too ambitious about how many wires you can stuff in one hole (block connectors, like bullets, are available in several sizes).

With any of these options, always insulate around your connection - especially when soldering, or you'll be leaving bare metal exposed. Remember that you'll probably be shoving all the wires up into the dark recesses of the under-dash area - by the time the wires are nice and kinked/squashed together, that tiny bit of protruding wire might just touch that bit of metal bodywork.

Alarm installation

Security

01 Disconnect the cable from the negative battery terminal, and move the cable away from the battery. This will probably wipe out your stereo settings, but it's better than having sparks flying and your new alarm chirping during installation

02 Decide where you're going to mount the alarm/siren. Choose somewhere not easily reached from underneath. Try the siren in position before deciding. It's also best to pick a location away from where you'll be adding fluids to the window washer reservoir, oil to the engine or coolant to the radiator. Loosely fit the alarm to the bracket, to help you decide how well it'll fit in your chosen spot, then take the alarm away

03 Mark the position of the mounting holes

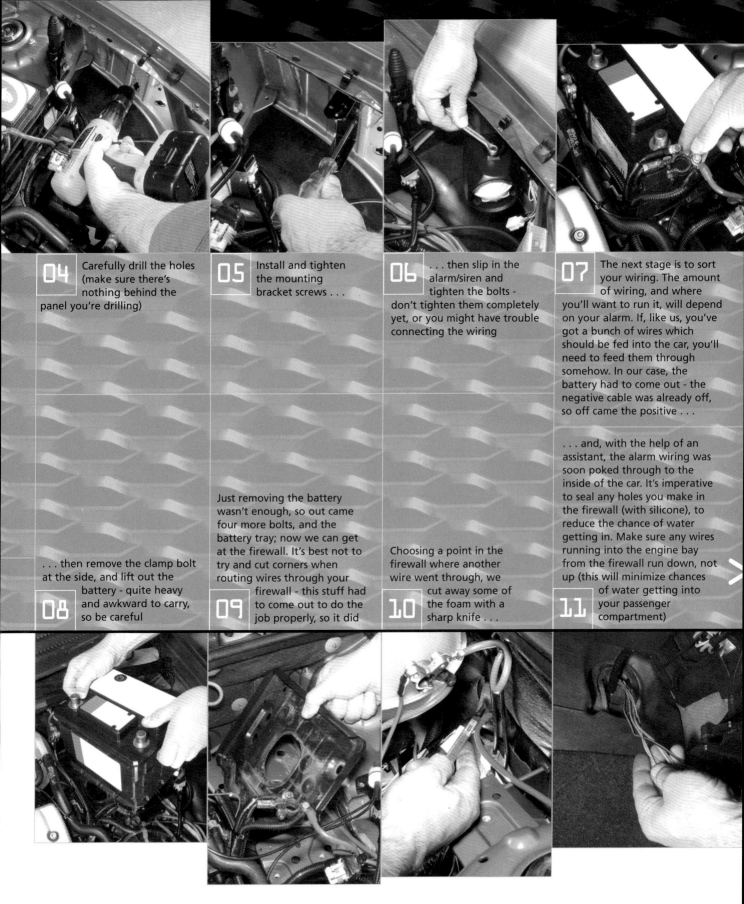

04 Carefully drill the holes (make sure there's nothing behind the panel you're drilling)

05 Install and tighten the mounting bracket screws . . .

06 . . . then slip in the alarm/siren and tighten the bolts - don't tighten them completely yet, or you might have trouble connecting the wiring

07 The next stage is to sort your wiring. The amount of wiring, and where you'll want to run it, will depend on your alarm. If, like us, you've got a bunch of wires which should be fed into the car, you'll need to feed them through somehow. In our case, the battery had to come out - the negative cable was already off, so off came the positive . . .

08 . . . then remove the clamp bolt at the side, and lift out the battery - quite heavy and awkward to carry, so be careful

09 Just removing the battery wasn't enough, so out came four more bolts, and the battery tray; now we can get at the firewall. It's best not to try and cut corners when routing wires through your firewall - this stuff had to come out to do the job properly, so it did

10 Choosing a point in the firewall where another wire went through, we cut away some of the foam with a sharp knife . . .

11 . . . and, with the help of an assistant, the alarm wiring was soon poked through to the inside of the car. It's imperative to seal any holes you make in the firewall (with silicone), to reduce the chance of water getting in. Make sure any wires running into the engine bay from the firewall run down, not up (this will minimize chances of water getting into your passenger compartment)

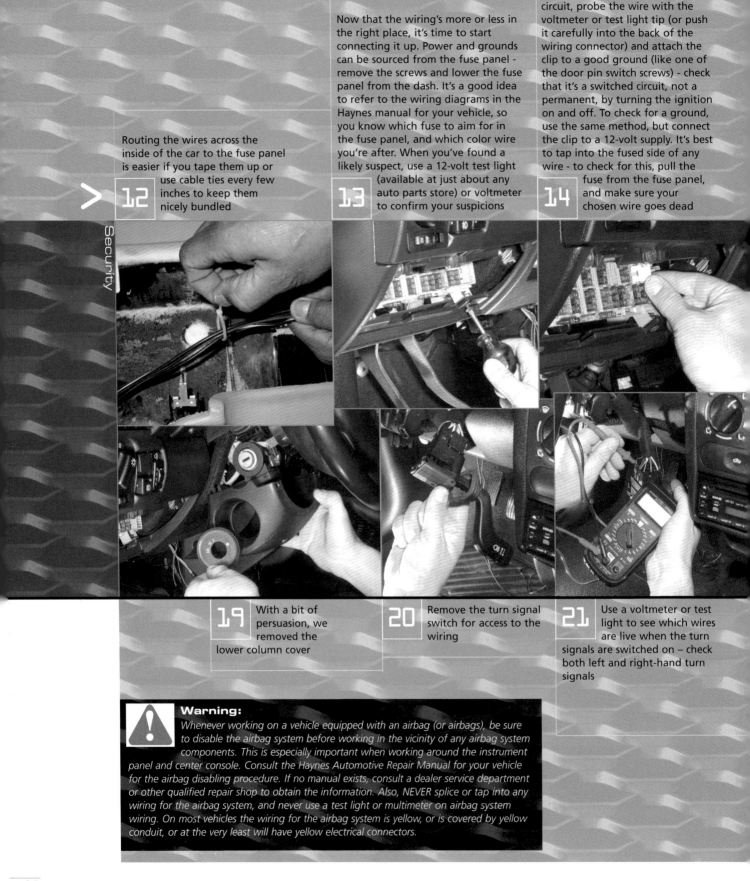

15 Most alarms are wired into the interior light circuit, which is operated by push-switches mounted in the door jambs. To get the wiring, unscrew the driver's door switch or remove the screw holding the switch in place, and pull the switch out

16 Disconnect the wiring plug from the switch, then use a test light to identify which of the two wires is live, and which is ground. Different alarms require you to wire into the interior light circuit on one wire or the other - check the instructions with your system

17 Most alarms require you to link into the turn signal circuit, so the lights flash during arming and disarming. One obvious place to tap into the indicators is at the turn signal switch, which means removing the steering column covers

18 Here we turned the wheel 90-degrees one way, then removed the screw. Turn the wheel back straight, then 90-degrees the other way, and removed the screw on the other side of the shrouds. Check your service manual for steering column cover removal, if necessary

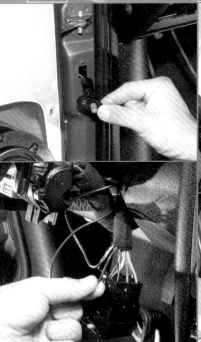

22 The best way to connect to any existing wiring without cutting it is to solder on your new alarm wires. It's permanent, won't come loose, and doesn't mess up the original circuit. Strip a little insulation off your target wire and the end of the alarm wire. Twist one around the other, if possible

23 Now bring in the soldering iron, heat the connection, and join the wires together with solder (be careful not to burn yourself, the dash, or the surrounding wires!)

24 Remember - whatever method you use for joining the new wires (and especially if you're soldering) - insulate the new connection big-time. The last thing you want is false alarms, other electrical problems, or even a fire, caused by poorly-insulated connections

19

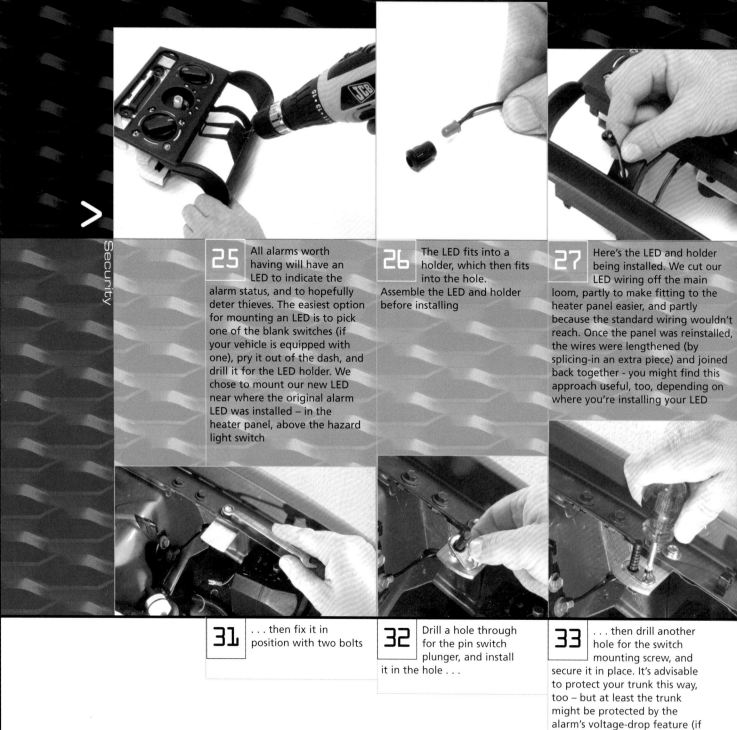

25 All alarms worth having will have an LED to indicate the alarm status, and to hopefully deter thieves. The easiest option for mounting an LED is to pick one of the blank switches (if your vehicle is equipped with one), pry it out of the dash, and drill it for the LED holder. We chose to mount our new LED near where the original alarm LED was installed – in the heater panel, above the hazard light switch

26 The LED fits into a holder, which then fits into the hole. Assemble the LED and holder before installing

27 Here's the LED and holder being installed. We cut our LED wiring off the main loom, partly to make fitting to the heater panel easier, and partly because the standard wiring wouldn't reach. Once the panel was reinstalled, the wires were lengthened (by splicing-in an extra piece) and joined back together - you might find this approach useful, too, depending on where you're installing your LED

31 . . . then fix it in position with two bolts

32 Drill a hole through for the pin switch plunger, and install it in the hole . . .

33 . . . then drill another hole for the switch mounting screw, and secure it in place. It's advisable to protect your trunk this way, too – but at least the trunk might be protected by the alarm's voltage-drop feature (if you've got a trunk light) or by the ultrasonics

28 You must protect your hood with a "pin switch." If a thief gets your hood open unhindered, he can then attack your alarm siren and any associated wiring. Game over. Install your pin switch close to the battery, to protect the battery connections. First, make a rough platform to mount your pin switch on, then hold a pen or scriber vertically on it and have an assistant slowly close the hood. Mark where the pin switch needs to be, to work, bearing in mind the "contours" of the hood

29 Your finished platform for the pin switch must be made of something pretty tough (metal seems obvious), otherwise it'll bend when the hood's shut. We just happened to have a nice thick piece of aluminum lying around, begging to be trimmed to size

30 Drilling holes in your fender isn't usually a good idea. Here, it's essential. Drill through the flange at the top of the inner fender, and through your plate . . .

34 The only things left to do now are to connect the wiring from your alarm, and then to test the switch operation. The switch should be set very sensitive – the alarm should go off, the instant the hood latch is pulled. The switch plunger is usually made of plastic and can be trimmed with a knife if necessary. If you need to lengthen the plunger, fit a self-tapping screw into the top

35 Now we're nearly there. Connect up the wiring plugs to the alarm/siren, and test it according to its instructions. Most require you to program the remotes before they'll work. Test all the alarm features in turn, remembering to allow enough time for the alarm to arm itself (usually about 30 seconds)

36 Set the anti-shock sensitivity with a thought to where you live and park - will it be set off every night by the neighbor's cat, or by kids playing football? When you're happy all is well, go around and tidy up the wiring with tape and cable-ties. There's a bit of a dilemma on the alarm fuses - if, like ours, yours are right next to the alarm module, do you tape them up, so a thief can't simply rip them out? Ah, but if you've buried them too well, you won't be able to install a new one so easily if it blows

04
Interiors

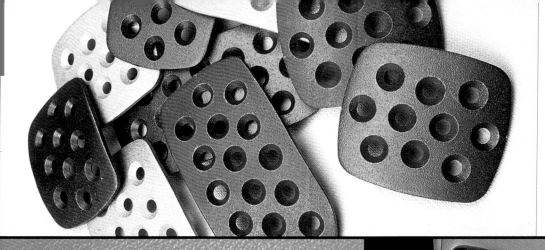

Are you sitting on an original equipment seat that bottomed out sometime back in the Nineties? Are you still wrapping your hands around a dull round blob of pebble-grain plastic that passes for a steering wheel? Are you looking at a set of gauges that are hard to read at night and don't tell you much anyway? Are you surrounded by a dull expanse of faded and scuffed plastic trim panels on the dash, console and doors? Well, lovers of style and color, rejoice! It's easy, and inexpensive, to do away with all this dullness.

There has never been such a wide range of colorful, stylish and attractive products for interior upgrades as there is today. From steering wheels to racing seats, from control pedals to trim panels, the only limits to what you can do to your car's interior are your budget and your creativity.

Sport
steering wheels

Look at your steering wheel. Dull, right? Wouldn't you rather wrap your hands around something really nice?

But first, a few words about airbags

If your pride and joy is *not* equipped with a driver's side airbag, you're in luck! You can install any steering wheel you want on your machine. But if your vehicle is equipped with a driver's side airbag we strongly recommend that you do NOT remove the airbag. First, and we hope this is obvious to you already, *it could save your life in an accident*. Second, *it's against the law*. Third, it's just dumb.

So what can you do if you have an airbag? Check out "styling rings" that install over the stock wheel, come in great colors and styles, and can give you a custom look without sacrificing your FACE in an accident.

Installing
an aftermarket steering wheel

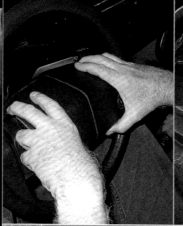

01 Position the steering wheel so that the wheels are pointing straight ahead. Then remove the horn pad. Look for a pair of bolts or screws on the front side of the steering wheel. These fasteners secure the horn pad. If you can't figure out how to remove the horn pad on your car, refer to the "Steering wheel - removal and installation" section in Chapter 10 of the Haynes manual for your car

02 After detaching the horn pad, disconnect the power and ground wires from the backside of the pad

Most steering wheels have either a "one-wire" or a "two-wire" horn button or horn pad. Two-wire systems have one wire for power and another wire for ground. One-wire systems have a power wire but use the metal parts (horn contact ring, steering wheel hub, etc.) as the ground path. Some aftermarket steering wheel kits are designed as a direct bolt-on replacement for your stock wheel, but most kits must be configured for a one-wire or a two-wire setup.

06

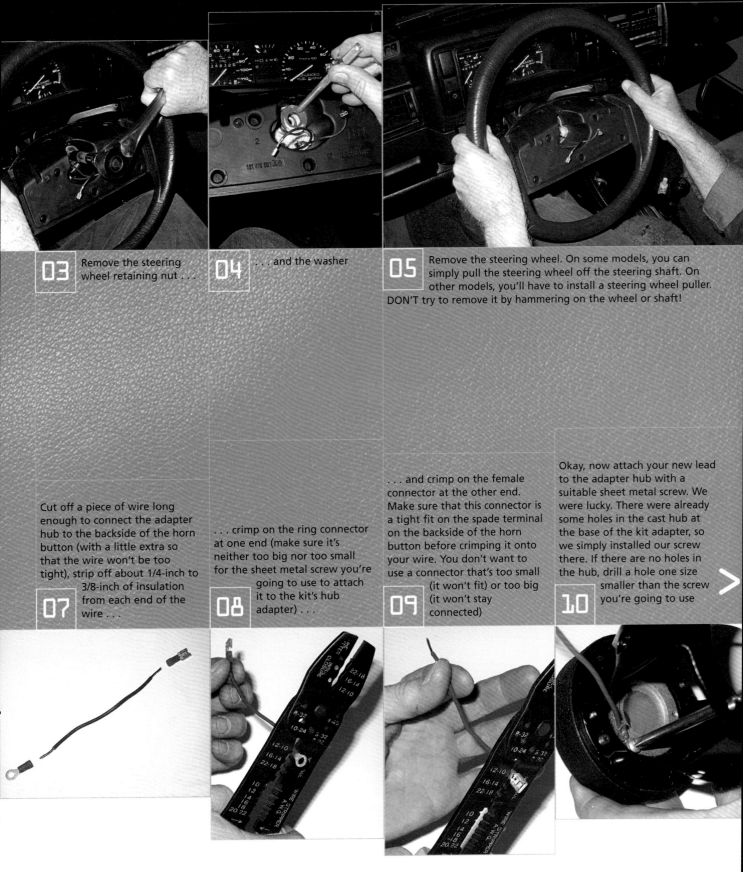

03 Remove the steering wheel retaining nut . . .

04 . . . and the washer

05 Remove the steering wheel. On some models, you can simply pull the steering wheel off the steering shaft. On other models, you'll have to install a steering wheel puller. DON'T try to remove it by hammering on the wheel or shaft!

07 Cut off a piece of wire long enough to connect the adapter hub to the backside of the horn button (with a little extra so that the wire won't be too tight), strip off about 1/4-inch to 3/8-inch of insulation from each end of the wire . . .

08 . . . crimp on the ring connector at one end (make sure it's neither too big nor too small for the sheet metal screw you're going to use to attach it to the kit's hub adapter) . . .

09 . . . and crimp on the female connector at the other end. Make sure that this connector is a tight fit on the spade terminal on the backside of the horn button before crimping it onto your wire. You don't want to use a connector that's too small (it won't fit) or too big (it won't stay connected)

10 Okay, now attach your new lead to the adapter hub with a suitable sheet metal screw. We were lucky. There were already some holes in the cast hub at the base of the kit adapter, so we simply installed our screw there. If there are no holes in the hub, drill a hole one size smaller than the screw you're going to use

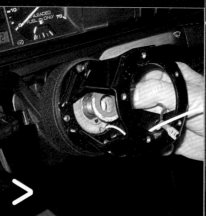   

11 Once you've got it wired, install the adapter hub on the steering column, install the washer . . .

12 . . . install the steering wheel retaining nut . . .

13 . . . and tighten it securely

14 Remove the horn button from the wheel. On our kit, the horn button is secured to the wheel by a series of locking tabs around the circumference of the button and the tabs are locked into place by a wire retainer

15 On our vehicle we had to cut off the portion of this wire contact spring that protrudes from the horn button assembly. (This is the contact for a one-wire system. On a two-wire system, if you don't cut it off, the horn will go on as soon as you reconnect the battery! Trust us. We tried installing the wheel without cutting it!)

16 Install the horn button in the steering wheel and secure it with the wire retainer. Then install this spacer and align the holes in the spacer with the holes in the steering wheel

17 Connect the power and ground wires to the terminals on the back of the horn button and install the steering wheel. Make sure that the spacer holes are aligned with the steering wheel holes

18 Install the steering wheel retaining bolts and tighten them securely. Reconnect the battery and try the horn button to verify that your electrical handiwork is good. That's it! You're done!

Have an airbag? Here's the solution

Installing a custom styling ring

1 Unpack your styling ring kit and then read the instructions

2 Clean off the steering wheel with a mild degreaser

Custom styling rings for airbag-equipped steering wheels

Even though you can't legally replace an airbag-equipped steering wheel with a custom aftermarket steering wheel, you can upgrade the appearance of your stock wheel with a custom styling ring. Each ring is custom molded to fit over the stock steering wheel. Styling rings are available in decorator colors like blue, red, silver, white, and yellow; in carbon fiber; and in simulated woodgrain such as burlwood or rosewood. They're easy to install and they can turn an ordinary airbag-equipped steering wheel into a stylish, elegant "new" steering wheel.

Carefully position the styling ring over the steering wheel. Before pressing it onto the steering wheel, make sure that the spoke covers on the ring are perfectly aligned with the spokes on the steering wheel. Then press the styling ring onto the steering wheel. Work your way around the circumference of the wheel, pressing down firmly all the way around to make sure that the styling ring is firmly attached. Easy, huh?

3 Clean off the inside of the styling ring . . .

4 . . . and then peel off the protective strip covering the adhesive

5

Gear shift knobs and boots

If you think how many times you actually use your shift knob and handbrake, surely, spending some hard earned money on how they look is well worth it.

Manual transaxles/transmissions

01 First, you have got to remove the old shift knob. On most vehicles this is done by unscrewing the shift knob from the shift lever

02 Use a small screwdriver to pry the plastic panel at the base of the shift boot

03 Pull the boot up and off the shift lever. The condition of this boot is worn and faded

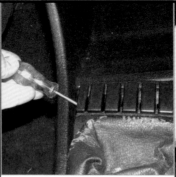

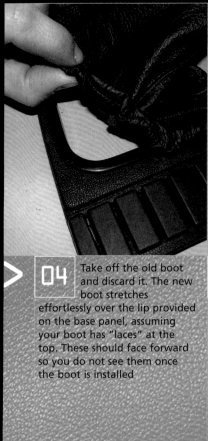

04 Take off the old boot and discard it. The new boot stretches effortlessly over the lip provided on the base panel, assuming your boot has "laces" at the top. These should face forward so you do not see them once the boot is installed

05 If you are going to install a trim ring around the base of the boot, slip it down over the boot and onto the lip

06 Use a screwdriver to tuck in any excess leather

07 Drill through the boot and the lip on the base panel to provide mounting holes for the screws. Try to keep the drill vertical while doing this or the mounting screws will enter off-set

08 Install the screws as straight as possible and tighten them with an Allen key

09 Install the new boot over the shift lever. Some boots have a threaded top section that attaches directly to the bottom of the shift knob

10 On this style shift knob, install the collar over the shift lever. Then, choose the tightest-fitting rubber cap to go over the threaded end

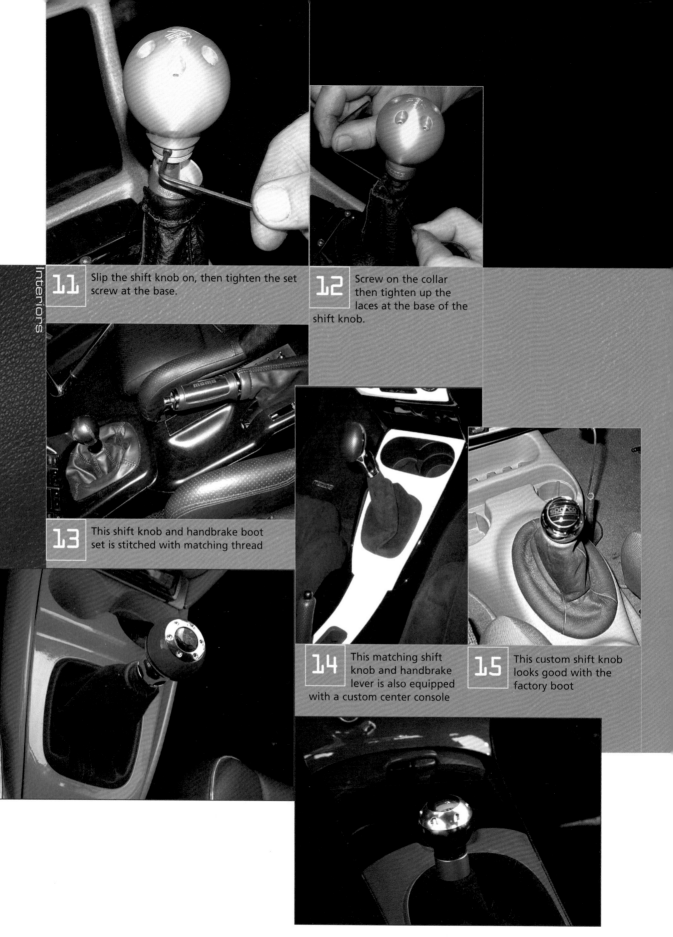

interiors

11 Slip the shift knob on, then tighten the set screw at the base.

12 Screw on the collar then tighten up the laces at the base of the shift knob.

13 This shift knob and handbrake boot set is stitched with matching thread

14 This matching shift knob and handbrake lever is also equipped with a custom center console

15 This custom shift knob looks good with the factory boot

01 This custom shift knob can easily be installed for the sport compact appearance on automatics

Automatic transaxles/ transmissions

02 Layout of the shift knob components

03 Remove the set screw from the factory shift knob

04 Some stock automatic shift knobs are equipped with two set screws

05 Press the shift control button and remove the knob from the shift lever

06 Determine which sleeve size (diameter) is correct for your shift lever

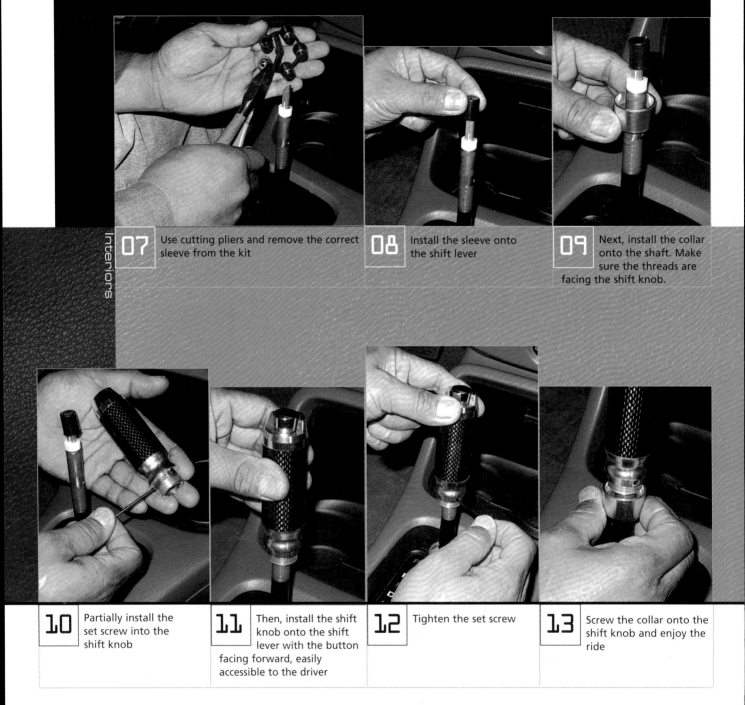

Interiors

07 Use cutting pliers and remove the correct sleeve from the kit

08 Install the sleeve onto the shift lever

09 Next, install the collar onto the shaft. Make sure the threads are facing the shift knob.

10 Partially install the set screw into the shift knob

11 Then, install the shift knob onto the shift lever with the button facing forward, easily accessible to the driver

12 Tighten the set screw

13 Screw the collar onto the shift knob and enjoy the ride

Handbrake boots

01 First unclip and remove the plastic cover

02 Use a tiny screwdriver to release the catch at the base of the handbrake

03 Slide the handle off the handbrake assembly. Some handles may have to be sliced to get them off the assembly!

04 Flip up the cover and remove the screw at the rear

05 Remove the lower cover also

06 This type of handbrake knob and boot kit requires three self-tapping sheet metal screws to bite into the handbrake lever. It may be necessary to purchase these specialized screws if they are not included with the kit. Turn the handbrake boot inside out to install the screws

07 After tightening all the screws, slide the boot over the screws. Check to make sure the handbrake button is working properly

08 Tuck the handbrake boot neatly inside the console. Install the rear cover screw (if equipped)

Custom pedals

Interiors

Have you ever noticed the accelerator, brake and clutch pedal pads on a racecar? They're not heavy rubber-covered steel pads like streetcar pedals. Instead, they're aluminum, with holes drilled in them for lightness. In other words, no frills - stripped for action!

You can pick a set of race replica pedals from literally hundreds of styles: die cast aluminum, brushed or polished, color-anodized, with or without lightening holes, with or without inserts. Some of the latest high-end pedals are now available in carbon fiber as well. The inserts (the small projections attached to the upper face of the pedal, to provide traction on the slippery surface of the pedal) are nylon, plastic, rubber or carbon fiber. Combined with other racy interior upgrades, a set of pedal pads gives your street machine a racecar look.

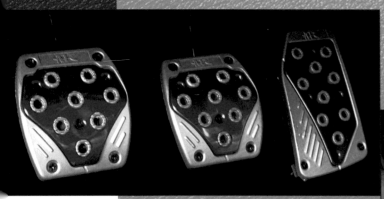

When selecting a set of race replica pedal pads, pay close attention to a couple of things. First, the new pedal pads will have three or four mounting bolt holes in them. When installing the new pedal pads, you'll be using these mounting holes to attach the new pads to the old steel pedals. But you'll have to drill mounting holes in the old steel pedals, and those holes *must be aligned with the mounting holes in the new pads*. So it's a good idea to either take the dimensions of your old steel pedals with you when you go to buy new pads, or to be able to take the new pads out to the parking lot, place them in position on your pedals and "eyeball" the dimensions. If the new pedal pads have mounting holes sitting over nothing but thin air when you position the new pads over the old pedals, think seriously about a different set of pedal pads! The existing steel pedals must be large enough so that you'll be able to drill mounting holes in them without relocating the pedal pads. Relocating the pedal pads could cause clearance problems between the new pedal pads. And, more importantly, it could be dangerous to offset the pedal pads because you might accidentally depress the wrong pedal at the wrong time.

The other thing to watch out for is the location of the pedal arm in relation to the mounting hole locations on the new pedal pads. If you're going to drill into the pedal arm when you drill any of the mounting bolt holes in the old steel pedal, read on to see what we had to do when we didn't consider this problem, then think about buying some other pedals! This interference problem with the pedal arm usually affects the *upper* mounting bolt hole on aftermarket pedals with mounting bolt holes in the *center* of the pedal.

01 First, remove the old rubber pads. On most cars, the stock rubber pads are held in place by a little adhesive. As the adhesive ages, it dries out, so you might be able to remove the old pads by giving them a firm yank! If necessary, use a sharp hobby knife to cut through any stubborn adhesive or rubber still bonded to the old steel pedal pad

02 Look at the mounting-bolt holes on each new pedal cover and note their locations in relation to the pedal arm. You're going to drill holes for the pedal cover mounting bolts in the factory pedals. If you find that the pedal arm is going to interfere with a mounting hole drilled through the pedal, DON'T try to clear the arm by offsetting the new pedal cover to the left or right. That could be dangerous. Instead, fix it as follows

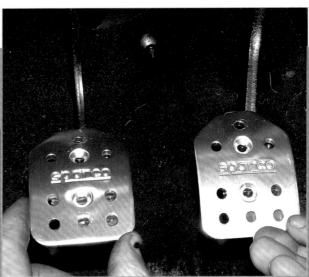

03 Note the mounting holes (the countersunk ones) on the clutch and brake pedals. Drilling a hole in the steel pedal pad directly below the upper mounting holes on these two pedals presents a problem because drilling here would mean drilling into the pedal arm

04 To solve the upper mounting bolt problem on the brake and clutch pedals, countersink the two holes on either side of the intended upper mounting bolt hole

05 With the mounting-bolt-hole problem solved, place each pedal cover on top of its respective pedal and drill the corresponding holes in the steel pedal

06 If necessary, support the steel pedal with a block of wood while you're drilling the mounting holes in the pedal. After you drill the first hole, install a mounting bolt and nut to secure the pad in its correct position while you drill the other holes

07 Once all the holes are drilled, remove the pedal covers and install the rubber plugs in the holes

08 Install the mounting bolts and tighten the nuts securely. (Since the brake and clutch pedal pads were designed to be installed with just two bolts, we had to find a couple of extra bolts locally)

09 If your pedal pad set uses countersunk Allen bolts (most kits do), hold the bolt with an Allen wrench and tighten the nut. If you strip out the recessed hex in the top of the bolt while trying to tighten the bolt with the Allen wrench, it'll be difficult to remove the pedal cover if it ever becomes necessary

10 Verify that all the pedals work correctly. Depress each pedal as far down as it will go. Make sure that none of the pedals catches the leading edge of the carpet or floormat. That's it! You're done. Enjoy!

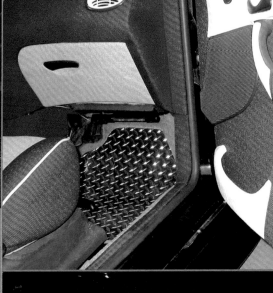

Custom floormats

Covering up worn and scrappy carpet can easily be accomplished using custom floormats. The types and patterns of floormats are endless, depending upon your style.

Interiors

Neon Lighting

Glowin' in the dark could not be easier or cheaper.
So switch the switch and get with it

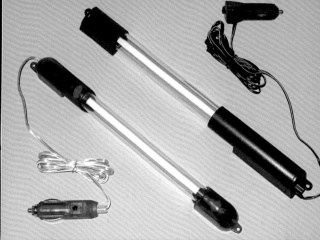

01 Most kits include an enclosed neon light assembly, an electrical lead, an adapter to plug into the cigarette lighter and the instructions. We decided to install a pair of purple neon lights under the dash of a Honda Civic, one for the driver and one for the front-seat passenger, and a blue neon light in the backseat footwell area

Installing neon lighting under the dash

02 Position the neon light assembly under the dash and try to find a place that will provide you with a solid installation. We picked this spot because it was flat and we could drill into it without hitting anything vital

03 Mark the location of the mounting screw holes

04 Drill the mounting screw holes with a drill bit a little bit smaller in diameter than the screws you're going to use

05 Install the mounting screws and tighten them securely, but don't really crank on them. This is a plastic accessory being mounted on a plastic dash, and we all know what happens to plastic when you overtighten it!

06 Either unplug your cigarette lighter or, if you have an electrical accessory receptacle (like the one we found on this late-model Honda Civic), use that

07 Plug in the electrical adapter. That's all there is to it for the passenger side. Installing the other neon light for the driver is virtually identical to this procedure, except for one thing . . .

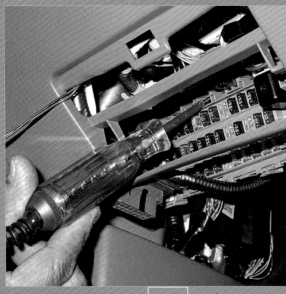

08 . . . if you install two (or more) neon lights, you can't plug them all into the cigarette light (or an accessory outlet). Of course, you could splice the two electrical leads into one adapter. But at that point you might as well splice the two (or more) leads through a switch you can flip on and off

⚠ Warning !
The use of neon lighting may not be legal in all areas. Check it out first. Also remember that driving at night with a brightly-lit interior makes it even harder to see out. Neons are best used at shows or in the parking lot.

Installing neon lighting in the rear footwell

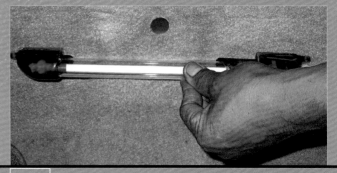

01 Position the neon light assembly where you want it. We selected the area right below the center of the back seat because we figured one light would provide plenty of illumination for both rear footwells

02 Using a laundry marker, mark the positions of the mounting screws

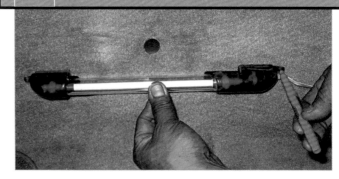

42

  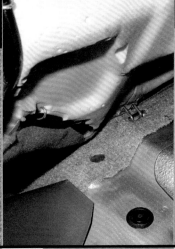 

03 We needed to figure out where to route the electrical lead for the neon light. We decided that behind and under the carpet was our best choice. So we removed the rear seat cushion bolts (see your Haynes manual for the specific rear seat removal procedure for your car) . . .

04 . . . then we lifted up the rear edge of the back seat cushion . . .

05 . . . unhooked the forward edge and removed the rear seat cushion

06 Then we removed this push fastener and peeled back the carpet and routed the neon light's electrical lead down and forward . . .

07 . . . and then pulled it through the hole in the carpet for the parking brake lever (center console removed). From here, we could route it forward, under the console, to the cigarette lighter or the switch we installed earlier for the front

08 With the electrical wiring routed under the carpet, place the carpet back in position and install the push fastener

09 Using the marks you made in Step 2, drill holes through the carpet and the body for the neon light mounting screws

10 Install the light, secure it with the mounting screws, reinstall the seat and you're done!

 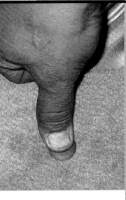  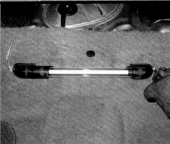

Interior trim
Add a little color to your interior

Ever hear that old saying about the Ford Model T? You could have any color you wanted as long as it was black. Well, automotive exteriors have come a long way since then, but most stock interiors are still mainly black (albeit many varied shades of black), or gray, or beige. Boring!

Painting trim

01 After removing the trim panels (a Haynes manual for your car would help here), use a degreaser to clean up the surface to be sprayed

02 Mask the areas you don't want to paint. Make sure that you protect all surfaces from overspray. Some paint manufacturers recommend putting the topcoat on very dusty, which means a lot of spray in the air. So take your time and mask it right. The extra time you spend will be repaid in the time you don't have to spend later, redoing the job!

03 Apply a mist coat of primer. This step is essential to help the paint stick to the plastic. Allow the primer plenty of time to dry

04 Apply the first topcoat very "dusty," which means you must spray from a little further away than normal, letting the paint *fall* onto the job, rather than blasting it on using the full force of the aerosol spray propellant. Some colors need several coats before they look right. Allow time for each one to dry (several minutes) before applying the next coat

05 Once you're happy with the coverage, let the last topcoat dry, then apply the sealer coat. Some sealer coats can give a glossier finish than you might prefer. So make sure you buy a sealer that's going to give you the kind of finish you want

06 Allow the paint to turn "tacky" (not wet, but not quite fully dry either) before you peel off the masking tape. And be careful when you remove the masking: If the paint's too dry, you might peel off some of the paint with the masking tape. (If you're now starting to feel a little paranoid, whip out your hobby knife and cut the edges of the masking tape before removing it!)

01 Using a suitable degreaser, clean up the surface you want to film. If you're hoping to film a heavily grained finish, be aware that the grain will show through thin film, and the film won't fully stick to a heavily grained surface either. (And don't try to remove the grain by grinding it off with Scotchbrite, or you'll ruin the panel surface)

02 Cut the film to the approximate size you need, but leave a generous amount of excess for trimming. It's also a good idea to leave plenty of film around the edges, because the film has a tendency to peel off otherwise

03 Carefully warm up the film and the panel itself with a heat gun. Peel off the backing sheet, and make SURE that the film stays as flat as possible. Also make sure that, when you pick up the film, you don't allow it to stick to itself!

Applying film

04 Carefully apply the film. If you're installing a patterned film (like the carbon-fiber-look film shown here), make very sure that you apply the film with the pattern aligned horizontally and vertically. Starting at one edge, work across to minimize air bubbles and creases. If you get a bad crease, unpeel the film a bit and try again. Don't try to shift the position of the film once you've begun to apply it; the adhesive is far too sticky

05 Before trimming your filmed panel work out the air bubbles with a soft cloth. Then make sure the film is sticking well by going over it firmly with the edge of a plastic credit card

06 Trimming the panel can be tricky. It's easier to trim the more complicated edges after heating up the film with a hairdryer or heat gun. But don't overdo it! Also, make sure that the hobby knife you're using is SHARP. A blunt knife will rip the film

07 To get the film to wrap neatly around a curved edge, make several slits almost up to the edge, then heat up the area you're working on, wrap each sliver of film over the edge and stick it on firmly. If you heat the film sufficiently, it wraps around and keeps its shape. Without the heat, the film might try and spring back, ruining all your hard work

Installing a custom trim kit

Unpack your new trim panel set and make sure that everything is there. And make sure that you have the right kit for your make and model! To keep the adhesive side of each individual trim piece dust-free, don't detach it from the backing paper until you're ready to install it. (And it's not a bad idea to wash your hands thoroughly to remove any oil so that you don't put fingerprints on anything!)

01 Here's the basic process. Let's start with the center console: First, wipe off all oil, dirt, silicone protectant, etc. with rubbing alcohol

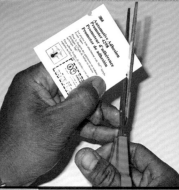

02 Then break out the promoter . . .

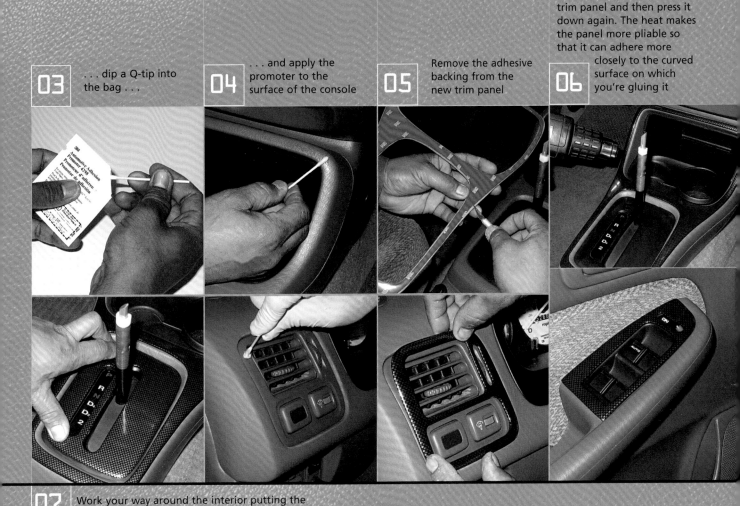

03 . . . dip a Q-tip into the bag . . .

04 . . . and apply the promoter to the surface of the console

05 Remove the adhesive backing from the new trim panel

06 Apply the new trim panel and push it down firmly. On curved surfaces like this console, use a heat gun to warm up the new trim panel and then press it down again. The heat makes the panel more pliable so that it can adhere more closely to the curved surface on which you're gluing it

07 Work your way around the interior putting the remaining pieces in place. Take your time

Interiors

Gauge face upgrade

White, colored, even glow-in-the-dark... changing out your stock black gauge faces will really add a custom touch to your dash

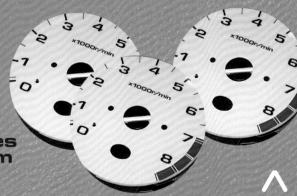

01 Unpack your electro-luminescent gauge face kit and make sure that everything is there. Then read through the instructions carefully before proceeding. Make sure that you understand the procedure. If you're nervous about any phase of this project, have a more experienced friend, preferably someone who's already installed luminescent gauge faces, help you. Or farm it out to a professional installer

02 The instrument cluster trim bezel for this Honda Civic is pretty typical of sport compacts. It has two upper retaining screws. The cluster trim bezel on most cars has at least two retaining screws and, depending on the make and model, as many as four screws, or maybe none at all. Refer to your Haynes manual if you're not sure

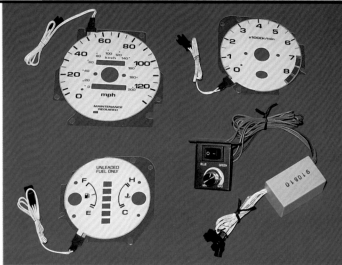

03 Using a short Phillips screwdriver, remove the instrument cluster trim bezel screws and then remove the bezel. If the bezel doesn't come right off, grab a penlight and inspect the entire perimeter of the bezel for "hidden" screws. On some vehicles, the screws are hidden under small plastic caps. Don't ever try to force a trim bezel loose or you will end up breaking off retaining tabs or the breaking the bezel itself. Ouch! These pieces are more expensive than they look

04 Once the cluster trim bezel is out of the way, find the (usually four) instrument cluster retaining screws (upper screws not visible in this photo) . . .

05 . . . remove the lower screws . . .

06 . . . remove the upper screws . . .

Just make sure you get the right kit for your car, and don't start stripping anything until you're sure it's the right one. Look carefully at every detail. Applications for replacement faces should be very specific for model and model year. This is one mod that is not ONE SIZE FITS ALL.

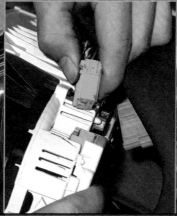

07 . . . pull out the instrument cluster far enough to unplug the electrical connectors from the back . . .

08 . . . and (at least on Honda Civics) the top as well . . .

09 . . . make sure that everything is disconnected and then carefully remove the cluster

10 Okay, put the cluster on a clean workspace and take it apart. Our cluster was held together by a bunch of little locking tabs. We used thin metal strips and inserted one into the slot at each locking tab to hold the tab down. Don't try to pry each of the locking tabs loose with a screwdriver; as soon as you let go of one tab and go to the next, the previous tab will snap back into place

11 Hey, it might look a little goofy, but it works!

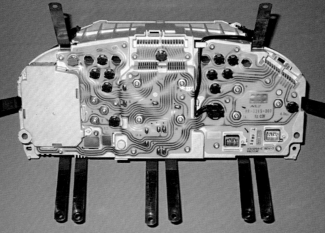

12 There's one more little thing you need to do before you can separate the two halves of the cluster assembly: remove any gauge illumination bulb holders with leads that cross the split line between the two halves of the cluster. Before proceeding, make sure that there are no wires or anything else routed between the two halves of the cluster

13 Using a screwdriver, very carefully pry the two halves of the cluster apart. Take your time and gently work the two halves apart by prying all the way around the perimeter of the cluster assembly. If you find an area that's difficult to separate, check the locking tab(s) in that area and make sure that it's still released

14 Separate the two halves of the cluster assembly and set the lens half aside

15 Remove the gauge faceplate from the cluster

16 Before further disassembly of the cluster, position each of the new gauge faces precisely over the old faces and make sure that the hole in the middle of each new gauge face is big enough to fit over the indicator needle hub. If it isn't, you've got the wrong kit for your car!

17 Remove the screws from the old tachometer gauge face and put them in a bag or some place where you won't lose them (they're small!)

18 Wash your hands! Make sure that there's no oil or grease on your fingers. Then very *carefully* work the new gauge face onto the tachometer by working the tach sweep needle up through the hole in the center of the new face and then sliding the face sideways until it's centered over the needle hub. Don't break off the tach needle or you'll be buying a new tach!

19 Holding the new tachometer face with one hand, turn the cluster over and look for a hole through which you can thread the electrical lead for the new faceplate. We lucked out and found this hole. We also found one for the fuel level and coolant temperature gauge, but had to make our own for the speedometer. If your cluster doesn't have a convenient hole for the tach lead, make one

20 Retrieve the old gauge face retaining screws and install them but don't tighten them yet. You're going to center the new gauge faces when you install the faceplate that surrounds the gauge faces. If you tighten the screws now, the new gauge faces won't be free to move

21 There was no hole for the electrical lead to the new speedometer face, so we drilled one in an area where we wouldn't hit anything expensive. Start with a smaller bit and then gradually work your way up to a 1/4-inch bit

22 Thread the new speedometer face electrical lead through the hole you made

23 Remove the retaining screws from the old speedometer face and stash them in the baggie or in someplace safe

24 Carefully slide the new speedometer face over the speedo needle and into position. Then grab your speedo face retaining screws and loosely install them

51

25 Look for a hole for the lead for the new fuel level/coolant temperature gauge face and thread it through the hole. If there is no hole, make one

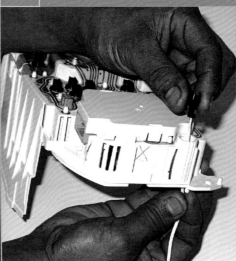

26 Remove the fuel level/coolant temperature gauge face retaining screws

27 Carefully install the new gauge face for the fuel level and coolant temperature. This one's a little tricky because there are two needles and hubs to deal with, but you'll figure it out! Just be careful and don't break off either of those needles!

28 Install the old gauge face retaining screws. Again, make sure that they're not tight yet

29 Place the faceplate in position

30 Make sure that the locator pins on the underside of the faceplate fit through the holes stamped through the transparent plastic edges of the new gauge faces. This centers each of the new gauge faces and locks them into position when the cluster is reassembled

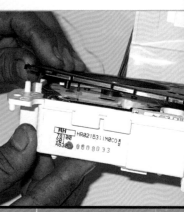

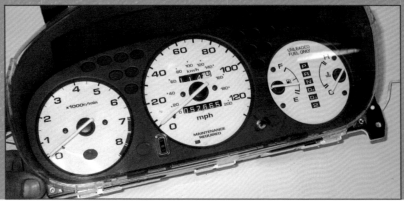

31 Now you can tighten the gauge face retaining screws! Just snug them. It's not necessary to really crank on them, or you'll strip out the threads

32 Holding the actual cluster in one hand, with the gauges and faceplate facing up, place the lens half of the cluster assembly in position and snap the two halves together. Make sure that all the locking tangs snap into place

33 Be sure to reinstall any illumination bulb holders you removed before disassembling the cluster and reroute the electrical lead through any guides or clips, just as it was before

34 Then combine the leads for the new new gauge faces into a small harness and bundle them together with two or three small cable ties

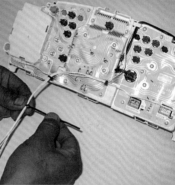

35 Okay, you're ready to install the cluster in the dash

36 Install the cluster mounting screws and tighten them securely. But don't install the cluster trim bezel just yet

37 Look for a good place to install the color selection switch and rheostat. We found a suitable location in the storage receptacle at the forward end of the center console

38 Mark the location of the holes you're going to drill for the switch mounting screws

53

39 Drill the mounting holes

40 Look for a place to route the switch wires. We inserted our wires through this gap between the upper and lower parts of the storage receptacle. If nothing is handy, drill a small hole nearby. Just keep in mind that you're going to route the wires toward the instrument cluster so that you can hook up with the three leads from the new gauge faces

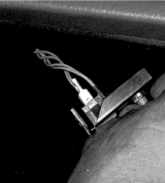

41 Install the switch/rheostat mounting screws and tighten them snugly but not so tight that you strip out the holes

42 Now fish out those three leads to the new gauge faces . . .

43 . . . and plug them into the leads from the switch/rheostat. (most kit switches will have three or four leads for this purpose, so you'll have an extra lead just in case you decide to reface another gauge later)

44 Now look for a terminal on the fuse panel that's hot when the parking lights are turned on

45 Strip off a 1/4-inch of insulation from the ground wire, crimp on a ground terminal and bolt the ground wire to a convenient metal-on-metal screw somewhere under the dash. Dash reinforcement bracket screws are good, and so are screws used to attach other ground wires. Just make sure that the ground screw makes a good connection to the body or floorpan

46 Strip off a 1/4-inch of insulation from the power wire, crimp on a suitable spade connector and connect the power wire to the hot-when-the-parking-lights-are-turned-on terminal at the fuse panel. That's it! You're done! Enjoy!

Four-point harnesses

Besides a roll cage, nothing says "racecar" like a four-point harness.

Until recently, in fact, the only place you could even see something this exotic was at the racetrack. Now, this serious safety component is finally available to street machine enthusiasts too. BUT, if you're thinking of installing a four-point harness, then you better really, really want one. Why? Because it can seriously reduce your ability to carry rear-seat passengers. Why? Because by the time you finish installing a pair of front-seat four-point harnesses, you'll have a webbing jungle in the backseat area.

They can also be a bit tricky to install and you certainly want them installed to meet all the safety standards. So, before going too far, do some serious research for your specific make and model. We're going to show a basic installation here, but your car could be quite a bit different. Just be sure to follow the instructions that will come with the harness you purchase.

Installing a four-point harness

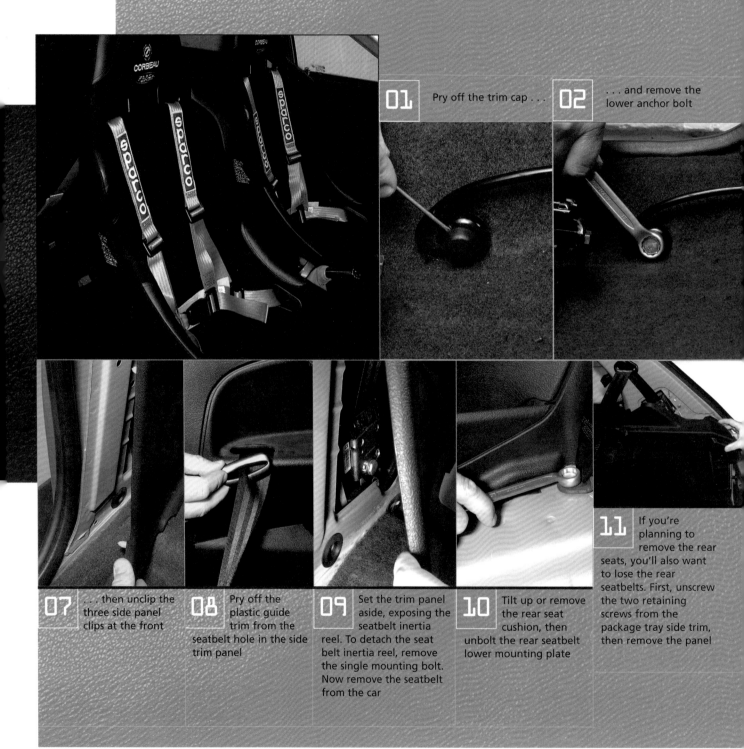

01 Pry off the trim cap...

02 ...and remove the lower anchor bolt

07 ...then unclip the three side panel clips at the front

08 Pry off the plastic guide trim from the seatbelt hole in the side trim panel

09 Set the trim panel aside, exposing the seatbelt inertia reel. To detach the seat belt inertia reel, remove the single mounting bolt. Now remove the seatbelt from the car

10 Tilt up or remove the rear seat cushion, then unbolt the rear seatbelt lower mounting plate

11 If you're planning to remove the rear seats, you'll also want to lose the rear seatbelts. First, unscrew the two retaining screws from the package tray side trim, then remove the panel

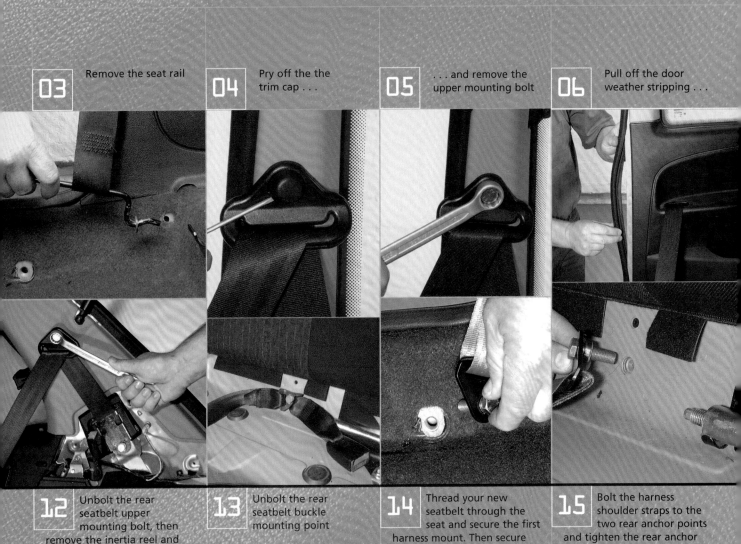

Bucket seats

If you're doing a complete makeover of the interior, sooner or later you'll have to decide what to do with the old seats.

If you've already discovered how difficult it is to brace yourself during hard cornering in a seat with no support, you're probably ready to replace the front seats with something a little sportier. And if your tired old La-Z-Boys just happen to be bottomed out, broken, stained or threadbare, then the decision's made! Make a pair of aftermarket bucket seats the centerpiece of your new interior. Bucket seats are available in a wide variety of colors and styles and features: Recliners and non-recliners; fabric, leather, vinyl, velour, suede (and pseudo-suede!); integral headrests and separate headrests; heated and non-heated; lumbar support and no lumbar support; and . . . well, you get the idea. Somewhere out there are two front seats with your name on them! But before you go seat shopping, here are some things to think about.

01 Move each front seat all the way forward to access the rear retaining bolts (you might have to remove small trim covers hiding the bolts) . . .

02 . . . then move the seat all the way to the rear to access the front retaining bolts. Tilt the seat upward and disconnect the electrical connector(s) for any power options (power seat motor, heated seat, adjustable lumbar support, etc.) . . .

03 . . . then remove the old seat and add it to your shop furniture!

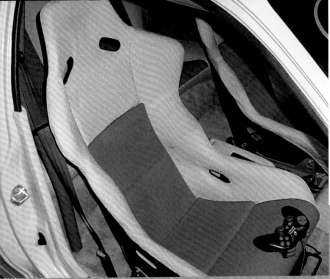

First, keep in mind that the seat mounting brackets and the "runners" (the rails on which the seat assembly slides back and forth) are *proprietary*. In other words, they're part of the original seat assembly, and a new seat won't necessarily fit without some modification. So make *sure* that there is a mounting kit for the seats you want so that you'll be able to install them in your car. If there isn't, then either think about some other seats, or be prepared to do some fabrication.

Original equipment seats may not be very supportive, but they are *adjustable fore-and-aft* and they can *recline*. Most aftermarket sport buckets are also adjustable fore-and-aft, and most of them are recliners. But if you're planning to install *real* racing seats, be forewarned: real racing seats are *not* recliners. They're *one-piece* buckets, which makes them stronger. If you decide to install non-recliners, the seatback angle will be fixed, so make *sure* that the seating position is comfortable for you.

Are you going to install a four-point harness with each new seat? If so, make sure that the new seats have holes in the headrest for the shoulder straps. Four-point harnesses *can* be installed in cars with seats that don't have these holes, but they won't look as integrated, or as racy. The good news is that you don't have to use (non-reclining) racing buckets to get those shoulder-strap holes. Many recliners are also equipped with these holes as well.

What we're showing here is a basic seat installation. All cars are a little different in the way the seats are mounted. A Haynes manual for your car will give detailed instructions. Also, be sure to follow the seat manufacturer's instructions and insure all safety-related guidelines are followed, i.e., don't disconnect airbag sensor wiring or factory seat belt components.

04 First, figure out which seat base is the left and which is the right. Then attach the seat bases to the new seats

05 The new buckets are now ready to install. Install the front and rear bolts loosely, then tighten them securely . . .

06 . . . and then torque them to the specification listed by the manufacturer

Starter button

Interiors

Real racecars don't use no stinkin' starter key. They use a starter button. And now, you can too! But be forewarned: if you're all thumbs when it comes to electrical work, have a professional do this job for you.

01 Disconnect the cable from the negative battery terminal

02 Remove the upper and lower steering column covers. To remove the ignition switch, insert a small screwdriver to release the lug . . .

Installing a starter button

07 To finish the wiring job at the ignition switch, strip some insulation from one of the big black wires we mentioned previously, and splice it to the ignition voltage supply wire to the relay. Again, solder the connection (not so easy on the thicker wire, but it'll happen if you're patient)

08 Clip the rear cover back onto the ignition switch

09 Then reconnect the switch plug to the ignition switch

03 . . . and separate the large wiring plug from the back of the ignition switch (you might have to remove a plastic cover to access the wires)

04 Identify the wiring colors for the ignition switch circuit (refer to the Wiring Diagrams in your Haynes manual). On our project vehicle, the wire to the starter solenoid is red/black and there were a couple of big black wires, either of which can be used to provide a live ignition supply voltage to the new relay we're going to install

05 Disconnect the red/black starter solenoid wire (or whatever color it is on your car) from the ignition switch wiring plug and then snip off the spade connector

06 Strip the end of the red/black wire and strip the new solenoid wire from the starter button kit (the blue wire in the photo). Connect the two wires and solder them, then insulate the connection with heat-shrink tubing or lots of electrical tape

10 Our kit's wiring included a relay (which you have to wire up) and a fuse holder. (If your kit doesn't include these two items, you'll have to obtain them separately.) Mount the fuse holder somewhere easy to reach so that you can change it without major disassembly. We tucked ours just under the bottom edge of the dash

11 We attached the fuse holder to the dash by drilling a hole and then screw the holder to the dash

12 To get to our fuse, all we'd have to do is remove the knee bolster trim panel. Easy!

13 To finish the wiring for the starter button, we need a ground (the voltage supply comes the "other side" of the relay). Ground wires are often brown, so we tested this plug's brown wires and found . . . ground! So we snipped it off at the plug . . .

>

14 . . . and connected it to one side of our new starter button switch. Once we'd also hooked up our black wire from the relay, we decided to give it a quick test - just to make sure that we'd wired it up correctly, before installing the button in the dash

15 The trickiest part of installing a starter button is figuring out where to put it! We picked the heater control panel area because it's centrally located, where it's easy to reach (and easy to show off!). We pried off the RECIRC slider knob from the lower heater panel . . .

16 . . . then pried off the lower panel itself. We thought drilling a hole in the right end of this panel would be a good plan

17 Put a block of wood behind the panel when drilling the hole

18 Insert the starter button in the hole . . .

19 . . . then turn the panel over and secure the starter button with the large locknut included in the kit

20 Back inside the car, hook up the wires to the starter switch terminals and tighten the set screws securely

21 Clip the panel back into position and install the RECIRC slider knob. That's it! What do you think?

  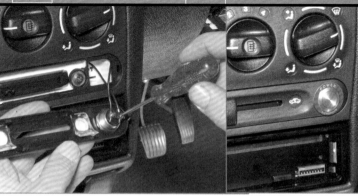

Installing aftermarket gauges

Installing pillar pod gauges

01 Get your pillar pod and dummy it up exactly where you want to put it. Make sure it's a good fit before proceeding. Then, drill a hole at each corner of the pod as shown

Interiors

02 Place the pod in position on the pillar and mark the locations of the four holes you're going to drill. Make sure the holes aren't going to be too close to the edges of the pillar

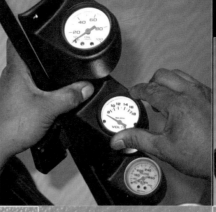

03 Insert the gauges into the pillar pod, place the pod in position on the pillar, then rotate the gauges so the "OIL" and "VOLTS" on the gauge faces are horizontal and parallel to each other.

04 Reinstall the gauges in the pillar pod and clamp them into place with the clamps provided by the manufacturer. Note how we spliced the two illumination bulb wires and the two ground wires together.

05 Route the wires and the oil line for the oil pressure gauge through the gap between the end of the dash and the A-pillar, then install the pillar pod/gauge assembly on the pillar and attach it with the four mounting screws. Don't overtighten the screws or you'll strip out the holes. When you're done with this phase, hide the wires with some convoluted tubing.

Installing aftermarket gauges

Installing a tachometer

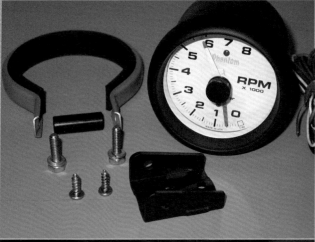

01 It's recommended to mount the tach where it's easy to see without taking your eyes off the road. Make sure that the dash material is substantial enough to support the tach. Once you've settled on the perfect spot for your tach, mark the position of the mounting bracket.

02 Remove the tach from the mounting bracket and mark the position of the bracket holes for drilling. Verify that your drill bit (and mounting screws) won't hit anything important like electrical wiring or vacuum lines

03 Install the tach in its mounting clamp, then bolt the clamp to the mounting bracket. Before tightening the mounting bolt and nut, hop in, adjust the seat to your regular driving position and adjust the angle of the tach so that it's facing directly at you

04 Hook up all the wiring exactly as instructed by the tach manufacturer, take your time for a clean, trouble-free installation

05 Unlike the other wires, the signal wire must be routed through the firewall to the engine compartment. Look for a convenient cable grommet (throttle, clutch, hood release, etc.) in the firewall and make a hole in it with an awl. We used the clutch cable grommet because it's big and because it's easy to get to

Window tinting

First, pick your day, and your working area, pretty carefully - on a windy day, there'll be more dust in the air, and it'll be a nightmare trying to stop the film flapping and folding onto itself while you're working. Applying window tint is best done on a warm day (or in a warm garage), because the adhesive will begin to dry sooner. Don't try tinting when it's starting to get dark!

The downside to tinting is that it will severely try your patience. If you're not a patient sort of person, this is one job which may well wind you up - you have been warned. Saying that, if you're calm and careful, and you follow the instructions to the letter, you could surprise yourself.

In brief, the process for tinting is to lay the film on the outside of the glass first, and cut it exactly to size. The protective layer is peeled off to expose the adhesive side, the film is transferred to the inside of the car (tricky) and then squeegeed into place (also tricky). All this must be done with scrupulous cleanliness, as any muck will ruin the effect (difficult if you're working outside). The other problem which won't surprise you is that getting rid of air bubbles and creases can take time. A long time. This is another test of patience, because if, as the instructions say, you've used plenty of spray, it will take a while to dry out and stick.

01 Step one is to get the window that will be tinted extra clean inside and out. Do not use glass cleaners or any other product using ammonia or vinegar, since both of these ingredients will react with the film tint or its adhesive and create a mess. It is also worth cleaning the working area around the windows because it is too easy for stray dirt to attach itself to the film tint. On door windows, lower them down partially to clean all of the top edge then close them tight to fit the film tint

02 Before you even unroll the film tint, beware - handle it carefully! If you crease it, you won't get the creases out. Unroll the film tint and cut it roughly to the size of the window

03 Spray the outside of the window with a weak, soapy water solution. Some film tint kits will provide a cleaning solution for your vehicle, but if not, use a little bit of dish soap in a spray bottle and apply the solution sparingly to the windows

04 Lay the sheet of tint onto the glass, with the protective film (liner) nearest you. Check this by applying a small piece of sticky tape to the backside and front side of the corners of the tint and film and carefully separate them. It will then become obvious which side is the sticky side of the tint and which side is the protective film

05 Spray the outside of the film with soapy water

Legal eagle:
The laws on window tinting vary from region to region and are sometimes confusing at best. Do some research on the film you intend to use and then contact your local authorities. Otherwise, install the tint with the understanding you could get stopped and have to strip it off.

Use a squeegee to get rid of the air bubbles and place the tint on the outside of the window glass. Remember, the protective film (liner) will be facing out

06

Use a sharp knife and be sure not to damage your paint or window rubber. Trim the perimeter of the tint to the outside of the window. On some rear glass and tailgate glass there are wide black bands on the edges of the glass. Cut your tint to the inside of these bands or the tint will not fit when it is transferred inside. Use a straight edge when cutting the tint

07

Now go inside the vehicle and prepare the glass for receiving the tint. Tape some plastic sheet to the door trim panel to prevent water damage when the tint is applied. It is a good idea to remove the door trim panel first, before going ahead with the job. Spray the inside of the glass with a soapy solution. Remember, no ammonia products such as glass cleaner or vinegar

08

Working on the outside of the glass, it is time to separate the tint from the protective film. Use two pieces of tape to pull apart the film at the corner

09

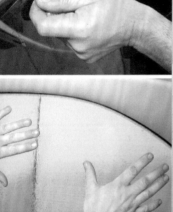

10 As the film comes apart, spray more solution on the tinted piece underneath to help it come apart cleanly. Try not to lift the tint off the glass too much as this will cause excess creasing. Have the assistant stabilize the tint to prevent any movement while the protective film is being removed from the tint layer

11 Have the assistant help transfer the tint from the outside of the window to the inside of the window. Peel the tint off the outside of the glass and keep it flat as possible. Remember, in this position, the outside tint layer contains the adhesive. Utilizing two people and without letting the tint fold, place the tint onto the inside window glass as close as possible to the correct position. The outside layer of tint now should be the adhesive side on the inside window. Carefully slide the tint into the corners, keeping the tint flat

12 Spray the tint with soapy water and carefully squeegee it into place, working from the top to the bottom. It is easier to use the squeegee blade separated from the handle to access the corner spots

13 You may end up with an area at the bottom of the glass that will not stick. Do not panic. First, soak up any excess water at the base of the tint with paper towels. Use a hot air gun to gently warm the tint at the base to assist with the adhesion. Be very careful when using a squeegee on a dry surface. Do not lift the tint off the glass. Be patient, the tint will stick. Persistence will pay off

05 Body & exterior

Aerodynamic/styling package, streamlined mirrors, trick taillights, de-badged decklid, custom paint - with the right combination of exterior modifications, you could have the coolest car on the block. And if you're really good at this game, you could start the Next Big Trend. Get it all wrong, however, and your precious project could end up looking more like some silly vehicle from a bad science fiction movie!

One way to avoid the latter outcome is to decide how you want to alter the appearance of your car and then do a lot of R & D: pore over catalogs and magazines, look at other guys' cars at shows, ask manufacturers the right questions before you buy anything. No matter what exterior look you're going for, the various pieces you'll need are out there. All you have to do is find them! It's a lot of fun, if you do it right.

Pay particularly close attention to "fit and finish." How well does a part - air dam, side skirt, rear valance, aero mirror, etc. - fit your car? Look for kits that fit well without requiring a lot of cutting, trimming or other alterations. How well finished is the kit? If it looks like #&*$ in gel coat, is it going to look any better painted? Be forewarned: body kits aren't all equal. Some fit the car they're designed for very well, some don't fit too well at all . . . even after considerable tinkering. Some are fiberglass, some are polyurethane (some of the newer kits are carbon fiber, but the prices of these kits will leave you gasping!).

De-badging

Metal emblems

Badges can be your best friend or your worst enemy.

If you've done some serious mods and want to go incognito; de-badge. If you want a nice, clean look; de-badge. And if you're tired of washing and waxing around all those chrome nameplates; take 'em off!

01 Our VW badge just pries off, but it leaves a deep recess behind. The Golf badge pries off too . . . best to do this with a screwdriver wrapped in rag to protect your paint.

02 This will leave behind little plugs which can also be pried out. Don't try too hard with this - some model badges are held in place by plastic tags which poke through the panel; take these out from behind with a sharp knife.

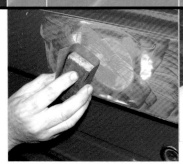

07 When it's dry, start rubbing down your filler first with coarse paper. If you're working on a flat surface, use a block with the paper wrapped around it, this way you'll be sanding consistently flat.

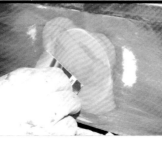

08 Then work down to something like 600 or even 1200-grade wet-and dry (best used wet, for this), and blend the filler out to the existing paint - see the "halo" effect we're starting to get? .

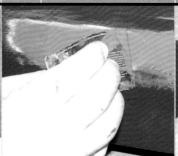

09 You'll probably need several light skimmings of filler before you fill in all the divots. Close your eyes and run your hand over the area, trust us, this way you're more likely to feel imperfections

10 Give the whole area a good wash before the next step. A pro would use cloths called "tack-rags" to pick up all the muck - see if you can get some. Spray on a light coat of primer first, spraying evenly from side to side in a criss-cross pattern is best. Then stop and notice all the lumps you didn't realize were in your smooth filler job.

Tricks 'n' tips

Don't stick the masking tape on completely flat. Stretch it out and stick down half the width, then curl the edge of the tape (nearest the new paint) over. You'll get a softer edge to the new paint, which will then be easier to blend in to the original. Sorted!

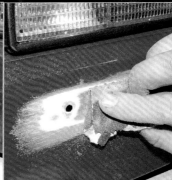

03 Since you're going be spraying at some stage, mask off (or even take off) your bumper now. Remember - you can never mask too much. Beware that anything less than proper masking paper can blot, which means the paint can soak through, so be careful. Being airborne means paint gets everywhere. Prep the holes with coarse sand- or emery-paper - this helps the filler to "bite".

04 Where our Golf badge was is a flat panel with three holes in it. If you try and fill these holes just as they are, the filler will fall out. Ideally you should weld these holes up, but as buying a welder especially for the job would be excessive, do this instead. Dish in the metal around the holes with a hammer . . .

05 . . . you'll see when you rub down if you've done this right.

06 There is only one way to apply filler, smooth it on in thin layers. Use a hole filling fiberglass based filler to start with, and apply a light skim of smooth filler when it's set.

>

Having found the lumps, get the block out again **11** and sand the primer down.

12 Now for the spraying. Build up in thin layers, giving each one a chance to dry - remember, you're not trying to cover in one coat (this is how runs occur). We gave our car a couple of coats of clear lacquer over the red, to give a decent shine and to stop it fading.

13 Carefully peel off the mask before the last coat dries completely - this will prevent the dried paint from cracking when the masking is removed

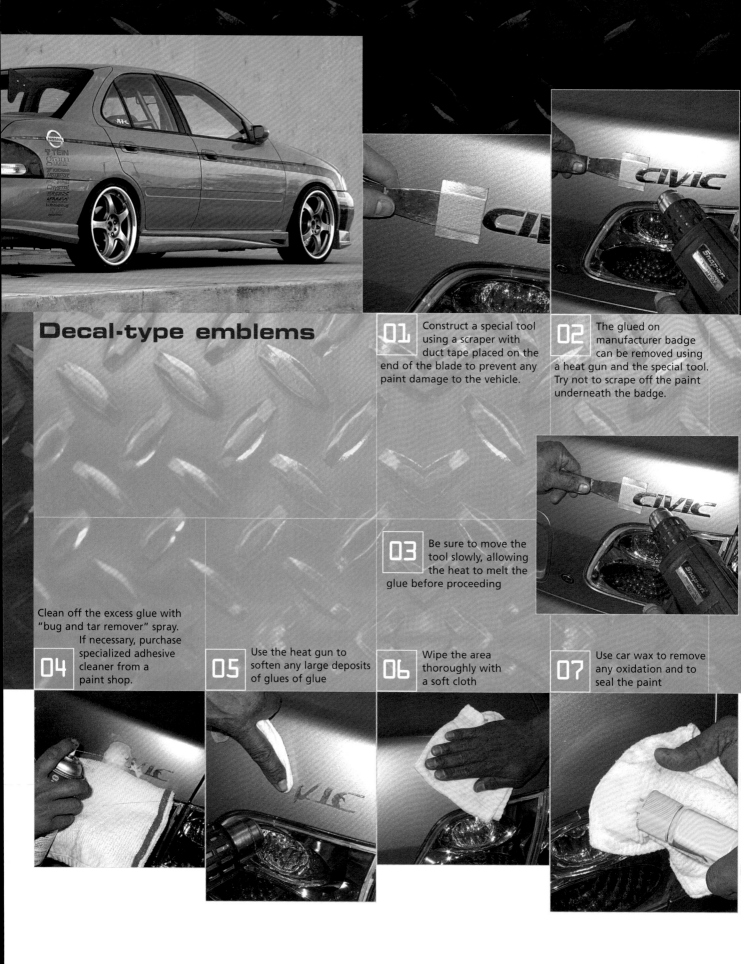

Decal-type emblems

01 Construct a special tool using a scraper with duct tape placed on the end of the blade to prevent any paint damage to the vehicle.

02 The glued on manufacturer badge can be removed using a heat gun and the special tool. Try not to scrape off the paint underneath the badge.

03 Be sure to move the tool slowly, allowing the heat to melt the glue before proceeding

04 Clean off the excess glue with "bug and tar remover" spray. If necessary, purchase specialized adhesive cleaner from a paint shop.

05 Use the heat gun to soften any large deposits of glues of glue

06 Wipe the area thoroughly with a soft cloth

07 Use car wax to remove any oxidation and to seal the paint

Aftermarket mirrors

You can install any mirror on any car, but it's not always easy. However, if you've got the skill, anything is possible.

Selecting the mirrors you want for your car is fun. When reading catalogs and car mags, look at the aftermarket mirrors installed on cars similar to yours. Car shows and racetrack parking lots are always good places to look for custom mirror installations. If you don't care for current styling trends, just be patient. Sooner or later, you will find some style that you really like. We liked the look of these "universal mount" mirrors so much that we obtained a set and installed them.

The manufacturer's claim of universality notwithstanding, the following procedure is by no means "universal." It will, however, give you some idea of what it takes to pull off a successful mirror swap. But please note: even if you decide to install a set of so-called universal-mount mirrors that's no guarantee that the swap will be "simple." For example, there's no easy way to convert a power mirror to a manually adjustable mirror. If you must have a truly hassle-free installation, then you should use mirrors designed specifically for your car.

>

Installing aftermarket mirrors

01 First, remove the mirror adjustment knob by pulling it straight off . . .

02 . . . then carefully pry the grommet from its hole in the door trim panel. Now start thinking about what you're going to do with this hole. Some kits come with a rubber or plastic plug to fill the hole or, if your kit has no plugs, start looking for a pair of rubber or plastic plugs that will fit. Or you could fabricate your own door trim panels, without the adjuster holes. Or - here comes the cheap solution - you could simply reinstall the stock pieces; they just won't do anything!

03 Carefully pry off the triangular plastic trim piece that hides the mirror mounting screws (it's attached to the door by a couple of plastic locator pins inserted through rubber grommets in the door metal) . . .

08 We found that the mirror mounting plates would fit neatly only after trimming the rubber window molding. Be extremely careful here, or you'll be ordering up new window molding!

09 Place each mirror in its installed position on its new mounting plate and mark the exact location of the mirror's mounting base on the mounting plate.

10 Determining where to drill the three holes for attaching the mirror to the new mounting plate isn't easy with this kit because the mirror mounting base isn't removable. If you have a similar problem, make a paper template: Place the mirror mounting base on a sheet of paper, mark the shape of the base on the paper, then cut out the paper and place it flush against the mirror base as shown. Using an awl or some other sharp object, "feel" for each screw hole and push the awl through the paper just enough to make a small hole (to ensure accuracy).

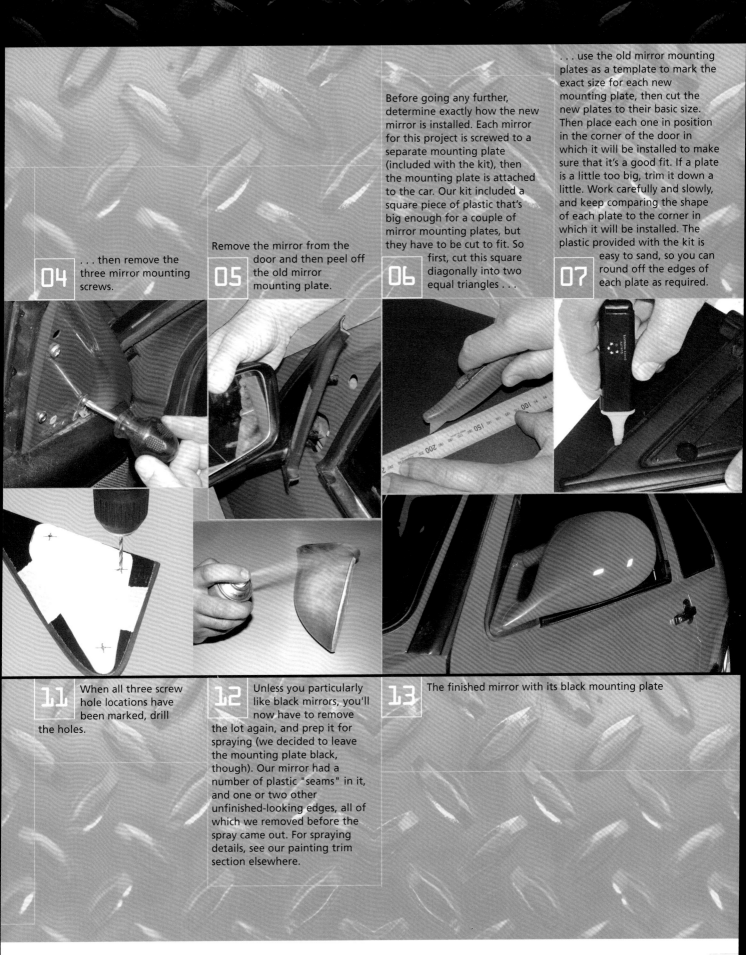

04 ...then remove the three mirror mounting screws.

05 Remove the mirror from the door and then peel off the old mirror mounting plate.

06 Before going any further, determine exactly how the new mirror is installed. Each mirror for this project is screwed to a separate mounting plate (included with the kit), then the mounting plate is attached to the car. Our kit included a square piece of plastic that's big enough for a couple of mirror mounting plates, but they have to be cut to fit. So first, cut this square diagonally into two equal triangles...

07 ...use the old mirror mounting plates as a template to mark the exact size for each new mounting plate, then cut the new plates to their basic size. Then place each one in position in the corner of the door in which it will be installed to make sure that it's a good fit. If a plate is a little too big, trim it down a little. Work carefully and slowly, and keep comparing the shape of each plate to the corner in which it will be installed. The plastic provided with the kit is easy to sand, so you can round off the edges of each plate as required.

11 When all three screw hole locations have been marked, drill the holes.

12 Unless you particularly like black mirrors, you'll now have to remove the lot again, and prep it for spraying (we decided to leave the mounting plate black, though). Our mirror had a number of plastic "seams" in it, and one or two other unfinished-looking edges, all of which we removed before the spray came out. For spraying details, see our painting trim section elsewhere.

13 The finished mirror with its black mounting plate

Color-matching exterior accessories

Sorry, this isn't the section in which we tell you how to spray-paint your entire car in a weekend. Nope. This section's about how to get that factory (yet custom) look to accessories you may add; like mirrors, taillight bezels, spoilers or wings.

01 Mask off any areas that you don't want to paint. Do this right at the start, or you could be sorry. On these door mirrors, we wanted to leave a black unpainted edge surrounding the mirror glass, so we decided to mask off the trailing edge of the mirror housing (and, of course, the mirror glass itself). If we had waited until later to mask this edge, we probably would have roughed up the shiny black plastic and wrecked the edge finish.

06 Another trick is to use a screw or, as shown here, a screw hole, to hang the piece you want to paint from some string or wire. Then you can spin the item around to get the spray into awkward areas.

07 Practice your spraying technique first. Working left to right, then right to left, press the nozzle so you start spraying just before you pass whatever you're spraying and follow through just past it on the other side. Keep the nozzle a *constant distance* (about six inches) - not a curved arc - from the surface being painted. Once you've got a patchy "mist coat" on (which might not even cover the whole thing), STOP! And then let it dry; primer dries pretty quickly. Continue building up thin coats until you've got full coverage, then let it dry for half an hour or more.

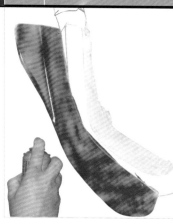

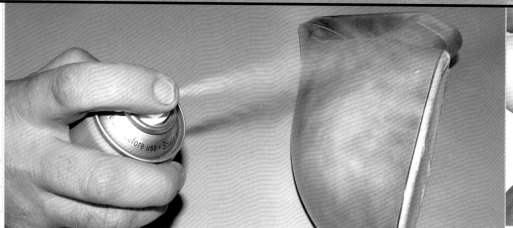

  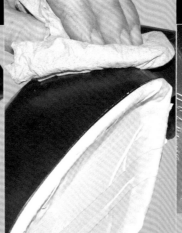 

02 Remove any unwanted "seams" in the plastic, using fine wet-or-dry sandpaper. Also tidy up any other areas you're not happy with, fit-wise, while you're at it.

03 Be sure to rough up the surface of plastic pieces before you spray them otherwise the paint will not adhere to the shiny or smooth surface. Just take off the shine, no more. You can use fine wet-and-dry paper for this, but Scotch-Brite is even better. This stuff is available from hardware stores and automotive paint and body supply shops. It's available in several grades. We used ultra-fine (gray color) on this plastic mirror housing.

04 Once the surface has been nicely roughed up, clean it off with a suitable degreaser (suitable means a type which won't dissolve plastic!). Generally, its okay to use rubbing alcohol or a cellulose-type thinner, but test it on a not-so-visible spot first to make sure that you don't experience an expensive meltdown!

05 Before you start spraying, it's a good idea to try a work a screw into one of the mounting holes, to use as a handle, so that you can turn the item to spray all sides.

Using 1000 or 1200-grade wet-or-dry paper, very lightly sand the entire primered surface to remove any minor imperfections (like paint glops, caused by "spitting" from the nozzle of the spray can). Try not to sand through the primer to the plastic (this doesn't matter so much in small areas). Use extra caution when sanding edges, because this is where it's easiest to sand through the primer. When you're done, rinse off the piece thoroughly and dry the surface.

08 Let it stand for a while to make sure it's completely dry.

Now it is time to squirt the color on. Give the can a good shake and let's get to it. Make sure that the surface is spotless. Build up a thin mist coat, allowing time for each pass to dry. You want to build a nice gloss without any runs. Any "dry" (dull) patches are usually caused by overspray landing on still-wet shiny paint; you can always polish these out afterward. Allow a couple of days for the paint to fully harden. Red fades easily, so apply a coat or two of clear lacquer over the color coat; this will also give you a glossy finish, even if you ended up with a dry finish. It's best to apply lacquer before the final color coat is fully hardened. The spraying technique is identical, although professionals say that

09 lacquer should be applied pretty thick - just watch out for runs! Lacquer also takes a good long while to dry. Pick up your newly painted piece too soon and you'll have that unique fingerprint effect!

 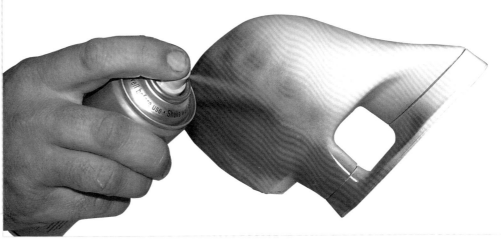

Body & exterior

Custom Taillight lenses

Available in lots of different styles, custom taillight lenses can be a quick, easy and inexpensive touch to your boring rear end.

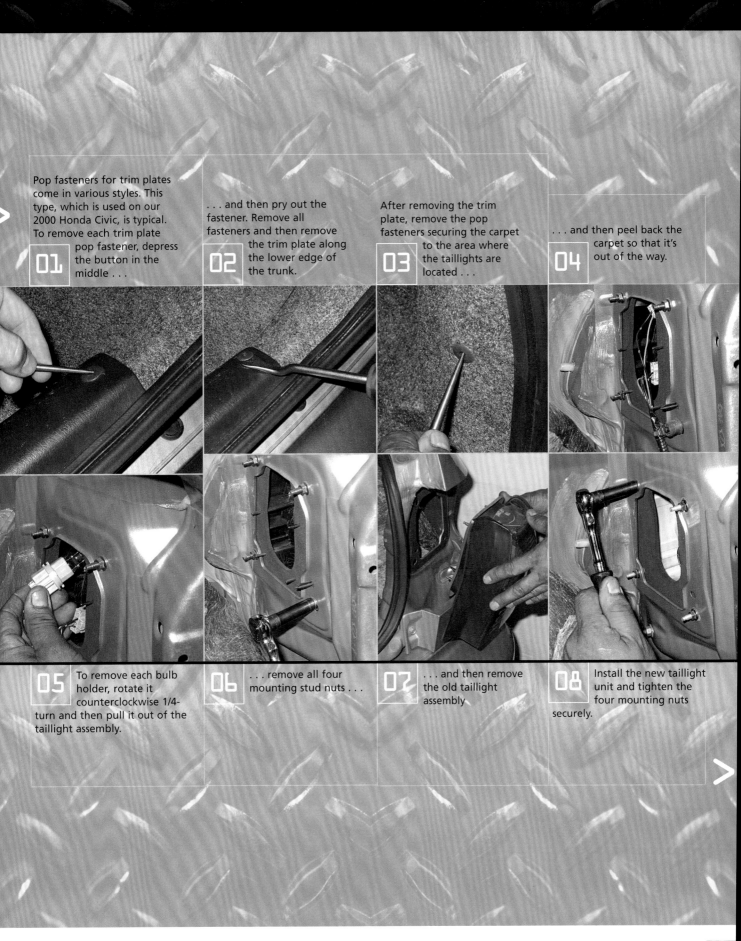

01 Pop fasteners for trim plates come in various styles. This type, which is used on our 2000 Honda Civic, is typical. To remove each trim plate pop fastener, depress the button in the middle . . .

02 . . . and then pry out the fastener. Remove all fasteners and then remove the trim plate along the lower edge of the trunk.

03 After removing the trim plate, remove the pop fasteners securing the carpet to the area where the taillights are located . . .

04 . . . and then peel back the carpet so that it's out of the way.

05 To remove each bulb holder, rotate it counterclockwise 1/4-turn and then pull it out of the taillight assembly.

06 . . . remove all four mounting stud nuts . . .

07 . . . and then remove the old taillight assembly

08 Install the new taillight unit and tighten the four mounting nuts securely.

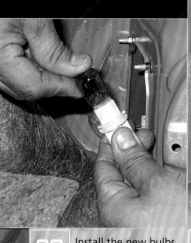 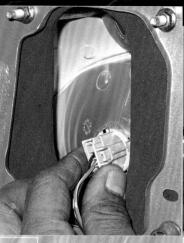  

09 Install the new bulbs in their holders (this is the amber bulb for the turn signal lights) . . .

10 . . . and then insert each bulb holder into its socket and give it a quarter-turn clockwise to lock it into place.

11 After both of the outer light assemblies have been installed, it's time to do the inner light units that are in the trunk lid. Remove the bulb holders and wiring

12 Remove all four taillight mounting nuts.

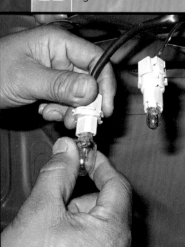 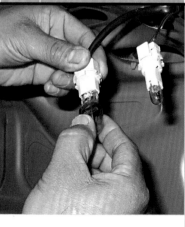 

13 Remove the clear bulb from the brake light bulb holder . . .

14 . . . and install the red light in its place.

15 Install the new taillight assembly in the deck lid

Install a custom grille

Nothing tidies up the front end of a project like a custom grille. You lose that dull stock grille, which is usually emblazoned with some goofy corporate logos. And in its place, you get a custom look which really sets your ride apart from the crowd.

And it's also functional because it allows even more cooling air to pass through the condenser, the radiator and the engine compartment. Until a few years ago, if you wanted a mesh grille on your car, you had to make it from scratch. Nowadays, a number of aftermarket manufacturers are offering bolt-on kits that can be easily installed in a matter of hours.

Body & exterior

01 Unpack you new mesh grille kit and make sure that everything, including all fasteners, is there. The kit that we purchased for a late-model Honda Civic includes everything you need to mesh the upper and lower grilles and the small faux brake vents in the bumper cover. Notice that it also includes an (unpainted) upper grille trim bezel, which must be primed and painted to match your car's color before installation.

02 Job One is to remove the stock upper grille and the bumper cover, which houses the lower grille and the faux brake vents; refer to your Haynes manual for the grille and bumper cover removal procedures.

03 Okay, once the stock upper grille and the bumper cover are removed, it's time to get started. First, turn the new grille trim bezel upside down and place the mesh grille in position on the bezel. Position the mesh as high as possible in the trim bezel to avoid any clearance problems with the bumper cover or the bumper cover fasteners.

07 Place the mesh grille trim bezel in place on the bumper cover, install the bezel mounting screws and tighten them securely.

08 Now grab the new lower mesh grille, place it in position and, using an awl, mark the locations of the mounting holes through the holes in the mounting tabs.

09 Put each faux brake vent grille in position and, using an awl, mark the positions of the two mounting studs. After drilling the holes, install the screws and tighten them securely

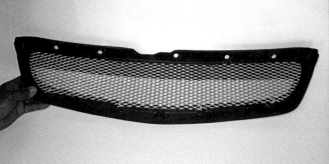

04 With the mesh grille centered in the trim bezel, drill holes in the bezel and secure the grille to the bezel with screws.

05 There you have it! Now remove the mesh grille so that you can paint the trim bezel (see *Color-matching exterior accessories*) or send it out to a professional to have it painted.

06 After the trim bezel has been painted and given enough time to dry thoroughly, reinstall the mesh grille.

Before...   **...After**

83

High Power headlight bulbs

Aftermarket headlight bulbs are an easy way to upgrade the power of your headlight beams.

This simple modification can give your old project vehicle the same lighting capabilities as the halogen and xenon units in new cars. However, please be aware that some of the bulbs out there might not be legal for use everywhere. Check before investing in bulbs that might earn you a ticket.

It's easy to install monster bulbs, but you should know about some of the typical issues that you might have to deal with if you do decide to install a pair of the ultra-high-power units. First, these big bulbs give off mass quantities of heat. Ask your friends; you may already know some people who have melted their headlight housings before they discovered this the hard way. You think we're exaggerating? Install a 100-watt headlight bulb, turn it on and put your hand in front of the light, next to the glass. The high heat generated by these bulbs will eventually damage the headlights, either by warping the reflector or by burning off the reflective coating. Or both!

The higher current needed to operate some of these bulbs has also been known to melt the wiring harness, which could, of course, lead to a fire, and which will almost certainly burn out your light switch. If there's no headlight relay installed by the manufacturer of your car, then it's safe to assume that the wiring and the switch were designed to carry no more current than the amount drawn by standard-wattage bulbs. If you're going for mega-high-power, install a relay to protect your light switch. (This is no more difficult than installing a relay for auxiliary driving light, fog lights or spotlights.)

01 First, pop the hood and see what's in the way. On this car, we had to remove this air intake duct (if you have an intake duct in the way, now would be an excellent time to run out and grab a cold-air intake kit!). These ducts are usually easy to remove because they're attached by one or two screws or plastic fasteners. To remove this duct, all we had to do was turn this fastener counterclockwise 90 degrees…

02 …then lift the duct up and out of the way. When removing this kind of stuff, be on the lookout for any wiring harnesses, vacuum lines or small devices such as airbag or engine management sensors or connectors that might be attached. If you have any problems trying to figure out how to remove something from the headlight area on your car, consult your Haynes repair manual.

03 At the back of the headlight, unplug the headlight electrical connector…

04 …then remove the rubber weather seal. (Not every weather seal looks exactly like this, but every headlight has one.)

The headlight bulb (or bulb holder) is usually secured by a wire retainer clip of some sort. To disengage a retainer like the one shown here (which is quite common), simply squeeze the sides together and swing it down…

05

06 …and then remove the bulb (or bulb holder). If your vehicle has a bulb holder (sort of a socket that the bulb plugs into), rotate the holder 1/4-turn counterclockwise and then pull it out of the headlight housing. To remove a bulb from a holder, rotate the bulb 1/4-turn counterclockwise and pull it straight out of the holder (some bulbs you simply pull straight out of the holder - no rotation is necessary). If you plan to reuse a stock bulb or give it to a friend, hold it only by the metal base, not the glass. Putting your oily fingerprints on the bulb itself will cause the bulb to darken and fail prematurely.

This "don't touch the glass" rule applies to your new bulb as well. If you accidentally touch the bulb glass, wipe it clean with Kleenex, and a little rubbing alcohol. Otherwise your new bulb will burn out much sooner than it's supposed to. Install the new bulb, secure it with the wire clip, install the rubber weatherproofing seal and plug in the headlight electrical connector. **07** That's it! Now go do the other side!

Body & exterior

Rear wing and spoilers

Let's be honest here. If you want to install a rear deck wing because you think it will actually improve the handling of your ride, think again. Unless you make regular trips at 130 mph+, you're not going to realize any real benefits. But if you want to install a rear wing because it looks good, at any speed (even standing still), well then just do it!

The number and variety of rear wings on the market nowadays is overwhelming. They're available in fiberglass, polyurethane, aluminum and, yes, even carbon fiber. They can be installed as part of a full body kit, or on their own. Paint to match the car or sleek black.

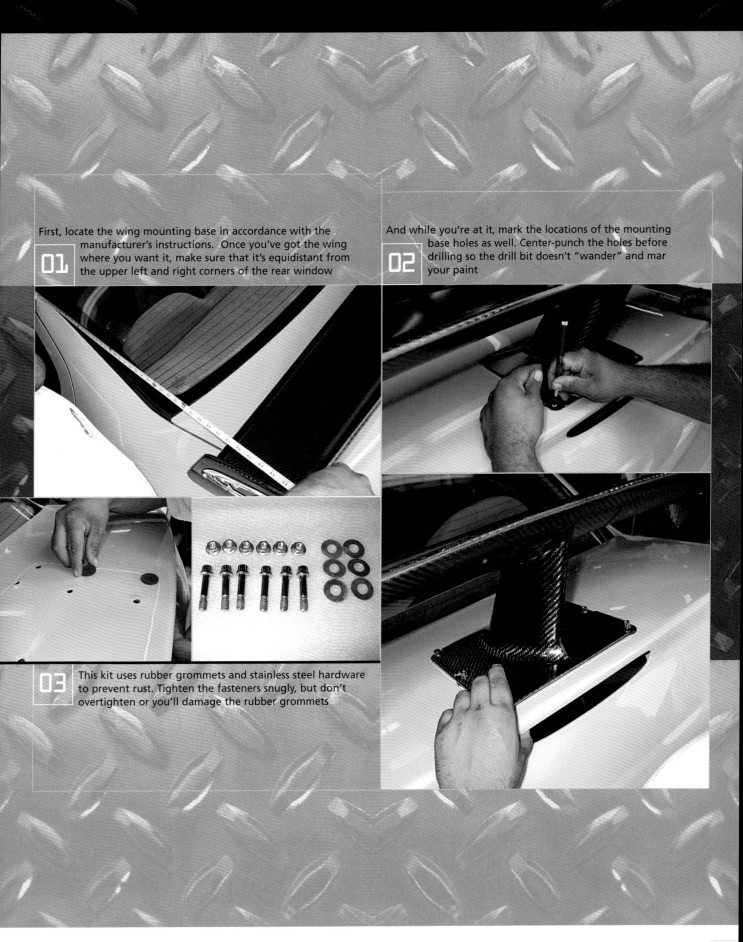

01 First, locate the wing mounting base in accordance with the manufacturer's instructions. Once you've got the wing where you want it, make sure that it's equidistant from the upper left and right corners of the rear window

02 And while you're at it, mark the locations of the mounting base holes as well. Center-punch the holes before drilling so the drill bit doesn't "wander" and mar your paint

03 This kit uses rubber grommets and stainless steel hardware to prevent rust. Tighten the fasteners snugly, but don't overtighten or you'll damage the rubber grommets

Body kits

Aside from a set of trick wheels or a custom paint job, an aerodynamic body kit is the biggest visual change you can make to your car.

The newest kits are easy to install because they already fit well, so little if any cutting or trimming is necessary. Some kits (like the Wings West prototype kit shown here for the Neon) are fiberglass, but others use high-quality polyurethane.

Most modern aero body kits come unpainted. And if you decide to purchase a polyurethane kit, it will be a little more difficult to paint than a fiberglass kit because you have to add a flex agent to the base coat. Nevertheless, most automotive painters say that polyurethane requires far less prep work than fiberglass. But, hey, we don't want to let all this talk about painting bore you, or scare you off. If painting's not your thing, fine. Some kit manufacturers will pre-paint their kits (for an additional charge) so that they match most factory colors (but not custom paint).

If you decide to go with an unpainted kit, you will of course have to measure, mark, drill, cut, trim and install some parts (like the rear bumper covers), then remove them for prepping and painting, then install them again. Other parts, like the multi-piece side skirts on this Neon, are pre-cut and no drilling is necessary. They're attached to the vehicle surface with double-face tape - really, really strong double-face tape, so they must be painted before installation. Once you stick those pieces on the car, they ain't goin' nowhere!

01 Unpack your new kit and familiarize yourself with all the pieces, where they go, how they're attached, etc. The kit instructions should include a list of all the parts. It's a good idea to inventory all the pieces and make sure that everything is there. By the way, if you're wondering why more fasteners weren't included, it's because these little parts don't use fasteners. You're going to stick them on with double-face tape. And the big pieces like the front and rear bumper covers actually use the existing fasteners (so save everything when you remove the stock stuff).

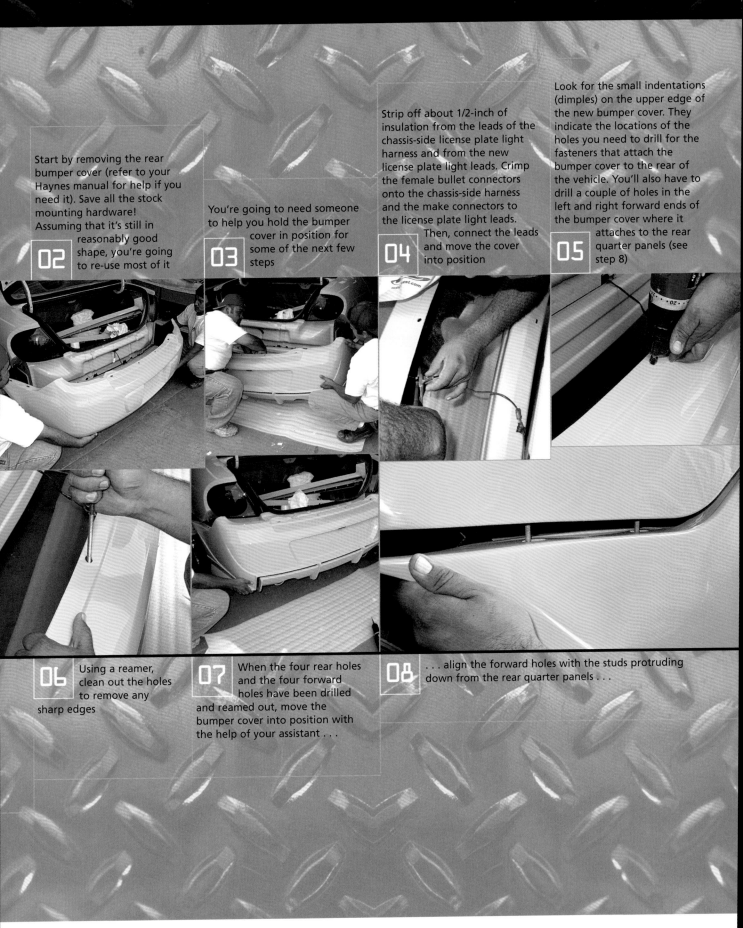

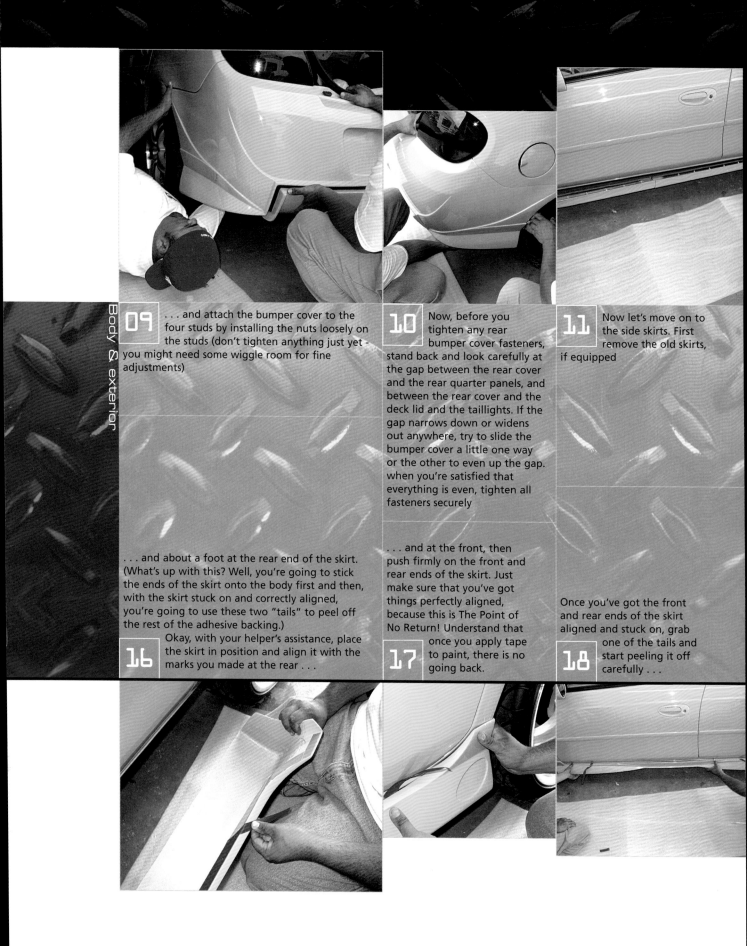

Body & exterior

09 ... and attach the bumper cover to the four studs by installing the nuts loosely on the studs (don't tighten anything just yet - you might need some wiggle room for fine adjustments)

10 Now, before you tighten any rear bumper cover fasteners, stand back and look carefully at the gap between the rear cover and the rear quarter panels, and between the rear cover and the deck lid and the taillights. If the gap narrows down or widens out anywhere, try to slide the bumper cover a little one way or the other to even up the gap. when you're satisfied that everything is even, tighten all fasteners securely

11 Now let's move on to the side skirts. First remove the old skirts, if equipped

... and about a foot at the rear end of the skirt. (What's up with this? Well, you're going to stick the ends of the skirt onto the body first and then, with the skirt stuck on and correctly aligned, you're going to use these two "tails" to peel off the rest of the adhesive backing.)

16 Okay, with your helper's assistance, place the skirt in position and align it with the marks you made at the rear ...

17 ... and at the front, then push firmly on the front and rear ends of the skirt. Just make sure that you've got things perfectly aligned, because this is The Point of No Return! Understand that once you apply tape to paint, there is no going back.

18 Once you've got the front and rear ends of the skirt aligned and stuck on, grab one of the tails and start peeling it off carefully ...

12 So grab the long skirt piece for one side of the car, place it in position exactly where you intend to install it and, with your helper holding the rear end securely in its correct position, use a china marker or grease pen to mark the upper edge of the piece from the front . . .

13 . . . to the rear. Then make vertical marks as well, to make sure that the piece doesn't move forward or backward during the actual installation

14 Using a grease remover or lacquer thinner, strip off all wax from the area where the side skirt's double face tape will come into contact. Then use a clean lint-free cloth to apply a thin and even film of adhesion promoter to the same area (Wings West recommends 3M No. 94 "Tape Primer" or Norton Performance Plastics "Tite-R-Bond"). Use the minimum amount that will fully coat the surface. When you're done, allow the promoter to dry completely before proceeding

15 When the promoter is dry, peel off about a foot of the adhesive backing from the 3M Double-Face Tape at the front end of the new skirt . . .

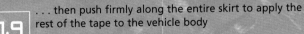
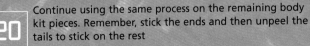

19 . . . then push firmly along the entire skirt to apply the rest of the tape to the vehicle body

20 Continue using the same process on the remaining body kit pieces. Remember, stick the ends and then unpeel the tails to stick on the rest

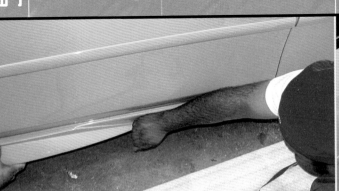

21 The old bumper cover has been removed and the new one is ready for installation

22 Heft the new bumper cover up into place . . .

23 . . . and secure it by loosely installing the nuts on the studs

24 Loosely install the three screws on top of the upper radiator crossmember

25 Okay, now check all your gaps and make sure that the new front end is correctly aligned, then tighten all the fasteners securely

Neon

Inexpensive and easy to install, neon lighting is one of the easiest ways to give your ride the "show car" look. With the variety of kits available today, you can install neon to just about any part of your car you want. Just try to keep it away from surfaces that could scrape the ground or get submerged easily in puddles. And check your local laws to be sure you're not doing anything illegal. When wiring, have a separate switch for your neon so it can be turned off when you don't "need" it, which will also help the components last longer.

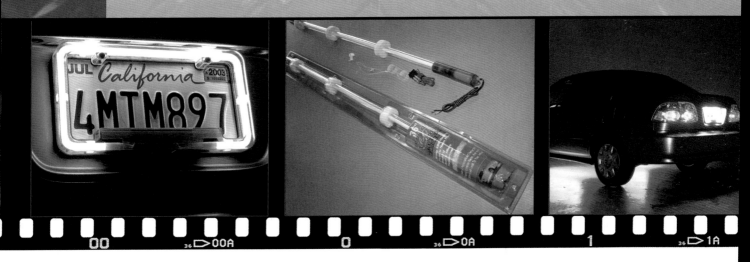

Body & exterior

Custom painting

Get some ideas of what you like – go to shows, look at books and magazines – and then find yourself a painter who can make it happen. If you can afford it, a custom paint job is the most effective way to make your car unique.

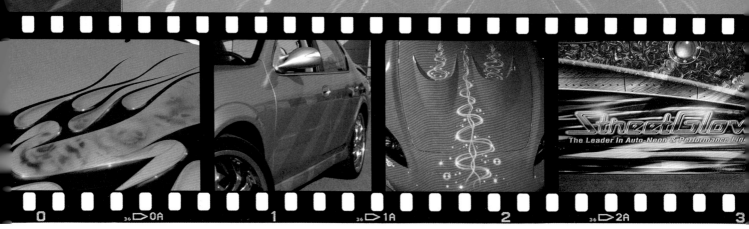

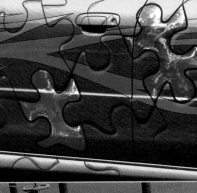

Metallic paints - have microscopic particles of metal in the paint that reflect the light and give off a high luster effect.

"Metal flake" paint - is really just metallic paint with bigger chunks of reflective metal in it.

Candy apple paint - consists of a reflective base coat of silver or gold metallic, with a translucent color coat on top of it and clear coat on top of that.

Pearlescent (or simply pearl) finishes - created by applying multiple layers of paint: first a matte color base, then a colored lacquer coat and finally a clear lacquer coat.

"Flip-flop" pearl or chameleon paint - use high-tech liquid crystal and interference pigments to produce a finish that looks like one color when viewed from some angles but looks like a different color when viewed from other angles.

06 Mobile entertainment

Mobile entertainment

DVD, Satellite radio, video screens in every headrest... isn't this a great world!

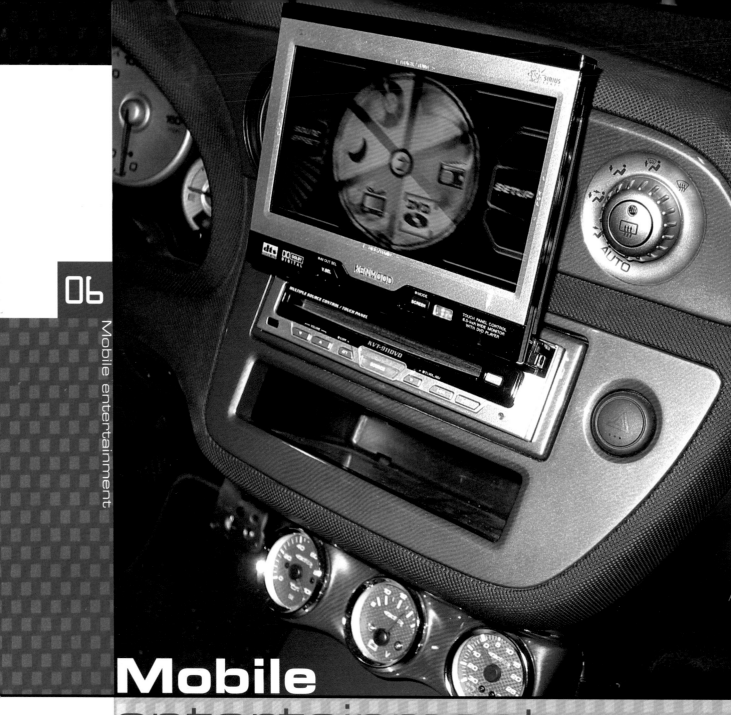

In-dash receivers & players

To give your audio system a decent start, you need a good head unit to provide the signal that an amp will beef up, and that the speakers will replay. Being the first link in the audio system's chain-of-command, makes it important to select the right head unit. You'd be well advised to spend some time selecting the right head unit for the job.

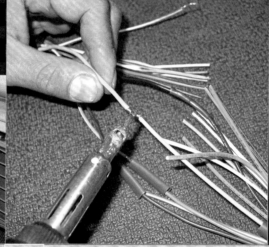

01 Refer to your vehicle's service manual on how to remove the factory radio. With this vehicle we had to first remove the dashboard trim panel . . .

02 . . . then by pushing the tabs at the sides of the radio, the unit was slid forward from its mounting bracket

03 Follow the manufacturer's instructions for connecting the wiring to the new radio. In this case we needed to solder an OEM connector to the radio's wiring harness (which will plug into the existing connector in the vehicle's dash)

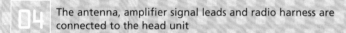

04 The antenna, amplifier signal leads and radio harness are connected to the head unit

05 Once it had clicked in, we added the dashboard trim panel to finish it off. Nice job!

Speakers

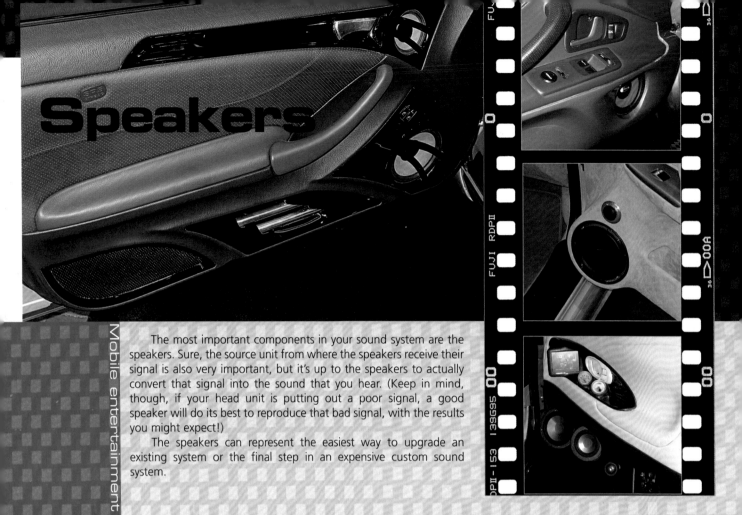

The most important components in your sound system are the speakers. Sure, the source unit from where the speakers receive their signal is also very important, but it's up to the speakers to actually convert that signal into the sound that you hear. (Keep in mind, though, if your head unit is putting out a poor signal, a good speaker will do its best to reproduce that bad signal, with the results you might expect!)

The speakers can represent the easiest way to upgrade an existing system or the final step in an expensive custom sound system.

Front door speakers and crossover

01 We chose to install a set of Focal Utopia component speakers with passive crossovers for the front doors. These are designed to be exact replacements for many vehicles

02 The door panel was pretty straightforward to remove, just be sure you've got all the screws out before you try to yank the panel off. Most door panels are also secured by push-in plastic fasteners around the perimeter of the panel (check your Haynes Automotive Repair Manual if necessary)

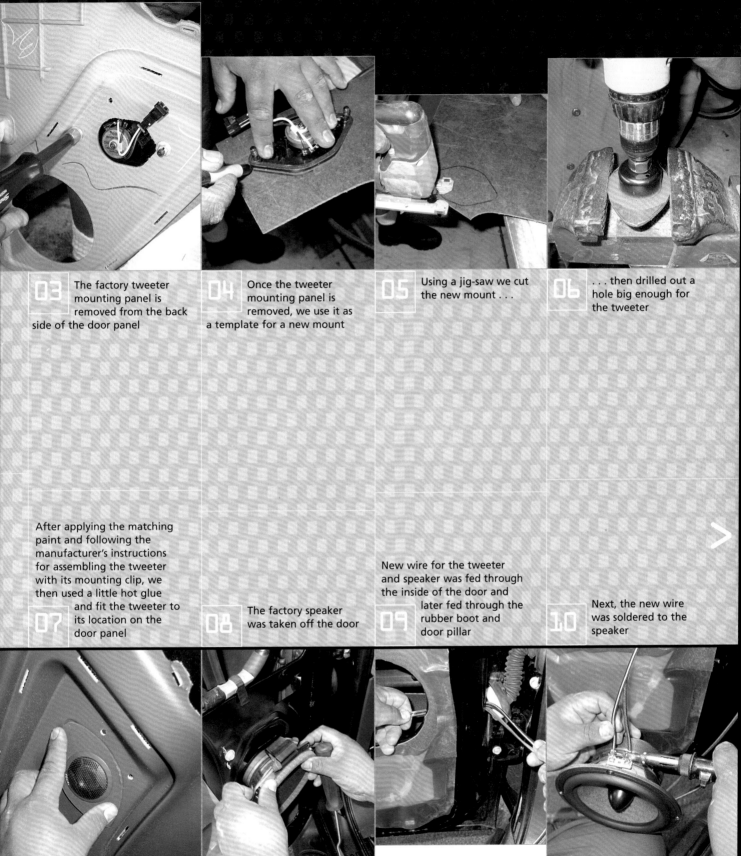

03 The factory tweeter mounting panel is removed from the back side of the door panel

04 Once the tweeter mounting panel is removed, we use it as a template for a new mount

05 Using a jig-saw we cut the new mount . . .

06 . . . then drilled out a hole big enough for the tweeter

07 After applying the matching paint and following the manufacturer's instructions for assembling the tweeter with its mounting clip, we then used a little hot glue and fit the tweeter to its location on the door panel

08 The factory speaker was taken off the door

09 New wire for the tweeter and speaker was fed through the inside of the door and later fed through the rubber boot and door pillar

10 Next, the new wire was soldered to the speaker

11. The speaker was then carefully mounted to the door

12. Back at the door pillar, we fed the tweeter and speaker wire through the harness rubber boot . . .

13. . . . then carefully reinstalled the boot to prevent any water leaks

14. After the door panel was installed, the tweeter wire needed extending to the crossover, so we soldered another length of wire onto the short lead coming from the tweeter. The wire was then fed under the carpet to where the crossovers were to be mounted

15. The crossovers were mounted conveniently next to the amplifier, under a rear seat

16. We added terminals to the wires . . .

17. . . . then following the manufacturer's instructions, connected them to the crossovers

Rear door speakers and crossovers

01 A set of Focal Polyglass coaxial speakers and passive crossovers are being installed in the rear doors. These are the direct replacements for the factory speakers and designed to complement the woofers/tweeters up front, and the sub-woofer in the back

02 We moved to the rear door and started by removing the door panel

03 The factory speaker was removed . . .

04 . . . and the crossover is connected to the speaker

05 Carefully the speaker is then mounted

06 By removing the door's harness boot . . .

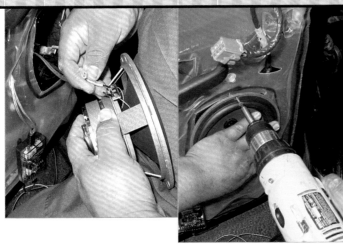

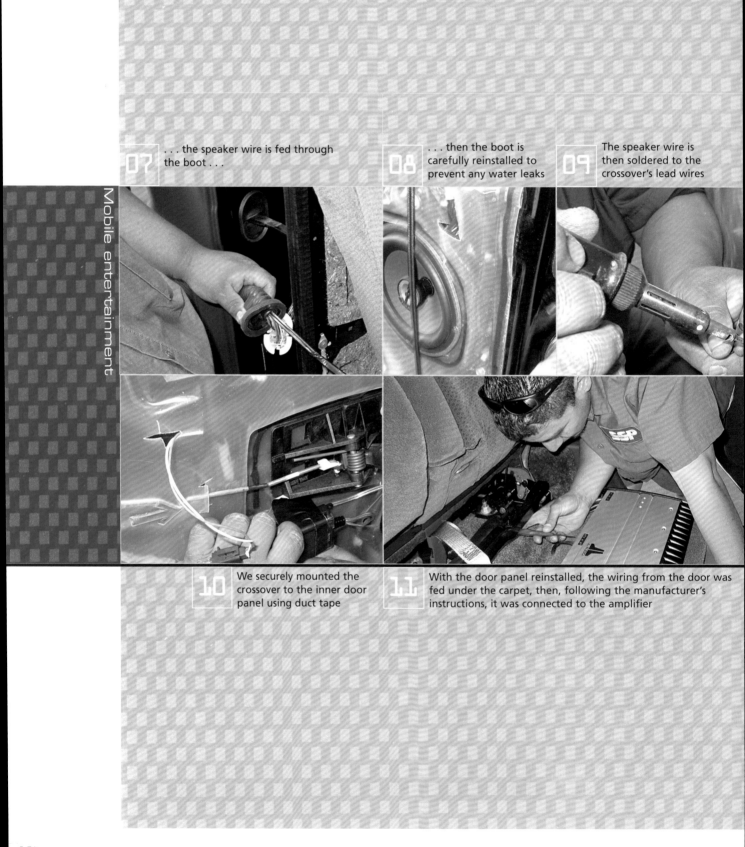

Mobile entertainment

07 ... the speaker wire is fed through the boot ...

08 ... then the boot is carefully reinstalled to prevent any water leaks

09 The speaker wire is then soldered to the crossover's lead wires

10 We securely mounted the crossover to the inner door panel using duct tape

11 With the door panel reinstalled, the wiring from the door was fed under the carpet, then, following the manufacturer's instructions, it was connected to the amplifier

We marked the speaker position on the underside of the shelf carefully to get both speakers in the same place

01

Using the speaker grille collar we drew in the position of the mounting screws and the outline of the opening...

02

Rear package tray speaker installation

The following procedure can be used to replace existing package tray speakers (without the fabrication of a new tray, like we're about to show here) or the addition of rear speakers in the package tray where none existed previously.

There are a couple of things that you should be aware of when using a medium density fiberboard (MDF) shelf, and the main one is the weight. They weigh a ton when they've got speakers mounted, so make sure you've got yours well secured before you take to the street.

 Warning: MDF dust is hazardous to your health. Wear a mask when you're cutting, drilling or sanding it.

03 ...and then drilled a few pilot holes in a row for the jigsaw blade. Also, drill pilot holes for the speaker screws so they'll twist in cleanly. After cutting the speaker holes we sprayed the underneath of the shelf black

04 With the shelf facing up, a couple of pieces of speaker cloth were pulled tight and stapled across the speaker holes. This was done to stop the new carpet trim from sagging into the speaker in the future

105

We stuck a little insulating tape over the staples to smooth them out before we started gluing the carpet and shelf together. We sized the carpet to make sure there was an adequate overlap under the shelf so that we could cut it off neatly. Then we stood the shelf on its edge on the upside-down carpet, ready for gluing. The glue was sprayed onto the shelf and the carpet, taking care to avoid the grille cloth that will cover the speakers. Once the glue had gotten tacky, the shelf was dropped forward onto the carpet and then flipped over to smooth out any wrinkles **05**

After masking up the shelf to give us some straight edges to work with, we glued the overlapping carpet and the shelf, then finished off the trimming **06**

The 6x9 speakers were screwed into the shelf using the pilot holes we'd drilled earlier, and then the wiring was added to them. Be careful you don't let the screwdriver slip off the screw and go through the cone, though! **07**

08 For the wiring hook-up we used push-on terminals for these speakers . . .

09 . . . and the cabling was tidied up using P clips to hold it at the front edge of the shelf. All that's left now is to mounting the shelf in the vehicle and connect the wiring (either to the factory wiring harness, if present, or to the head unit, crossover or amp, whatever the case may be)

Amplifiers

A good sound system needs power, and generally speaking, the more the better.

01 Deciding where the amplifier is going to be installed is the first step. In this case we're installing it in a small pick-up truck, so we've opted for mounting it under the front seat

02 Most of the time, removing the seat is a simple matter of removing the four bolts securing the seat to the floorpan and lifting the seat from the vehicle. You may also have to disconnect an electrical connector or two. If your vehicle is equipped with side-impact airbags, you'll have to disable the airbag system (refer to the Haynes Automotive Repair Manual for your vehicle)

03 Running the amplifier's power wire starts at the battery. The main power wire needs to have a waterproof fuse holder mounted as close to the battery as possible

04 Use a short piece of power wire between the battery and the fuse holder

05 Attach the other half of the fuse holder to one end of the short power wire . . .

06 . . . and crimp a terminal ring to the side that will connect to the battery

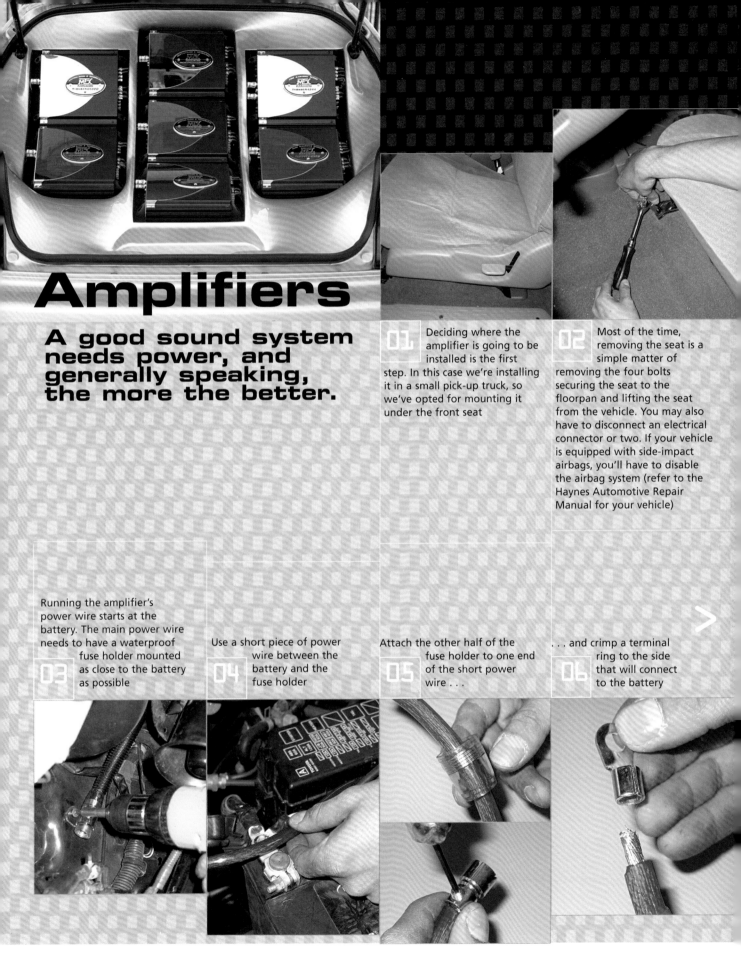

07 Connecting the power wire at the terminal is ok, just be sure NOT to install a fuse until you complete the entire installation

08 The power wire now needs to be routed to the passenger compartment. How to get the wire through the firewall doesn't have to be a dilemma, just find an existing hole like this one . . .

09 . . . remove the grommet, cut a small hole in it and feed the power wire through grommet and firewall. Be sure to refit the grommet to the firewall to prevent water leaks and to protect the wire

10 Inside the engine compartment, use convoluted tubing to cover and protect the entire power wire

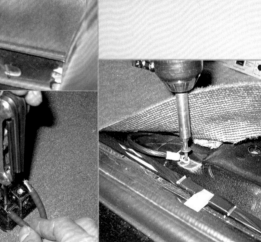

14 Route the amp's power wire and remote turn on wire together under the carpet - use wire ties or electrical tape to keep them tidy

15 A power distribution block can be used if two amplifiers are being installed

16 Cut a small piece of wire for the ground and crimp a ring terminal on one end. Find a spot to mount it to the vehicle chassis. Be sure to sand off any paint so that the connection is made directly to metal

17 Connect the power cable, remote turn on wire and ground wire to the amplifier. With this amplifier from JL Audio, the wires connect directly to the amp without the need for terminals

11 Inside the passenger compartment, properly secure the power wire to prevent being tangled or getting in the way of the pedals

12 The carpet is pulled back and the wire is directed towards the amplifier's mounting location

13 The amplifier's remote turn on wire and signal patch cable need to be connected at the back of the stereo head. Route the remote turn on wire in the direction of the amplifier's power wire where it enters at the firewall. *Note: When routing the signal patch cables, keep them a minimum 18 inches from any power wire including any vehicle wiring harness*

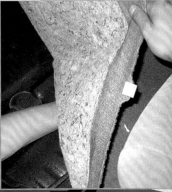

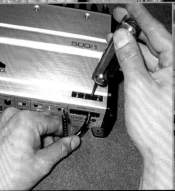

18 The same is done for the subwoofer speaker wires

19 Connect the signal patch cables to the amplifier

20 With the installation complete it's time to install the fuse and test the amplifier. Carefully follow the manufacturer's instructions for powering up the amp and making any necessary adjustments

 Note:
Always follow the manufacturer's recommendations for mounting the amplifier. Properly securing an amp is very important so that it's not sliding around. A sliding amplifier that's not properly mounted can damage the unit, or worse, be dangerous during an accident. The last thing you want during an accident is a UFO (unnecessary flying object) inside the passenger compartment.

Mobile entertainment

Subwoofer

As anyone who has attended a high-energy live concert can attest, there is no substitute for being able to *feel* the music. The addition of a subwoofer (or more than one) to bring out the low frequencies will add another dimension to your sound system and allow you to experience the performer's work with all of the "audio horsepower" that originally went into it!

Subwoofers are usually sold as a stand-alone item, but in just about all applications they will have to be mounted in some type of enclosure. These are big, heavy speakers that just can't be tossed into a door panel or under the dash. There are many types of enclosures, designed to manipulate the acoustics of the subwoofer(s) depending on the type of vehicle in which it is being installed or the type of music that will be listened to mostly. The physics behind these various designs is not easy to understand and is way beyond the scope of this manual, so we won't go there. However, your basic choices are to use a ready-made cabinet, an enclosure that has been designed to replace a center console or side panel specifically for your model car, or to custom build an enclosure yourself.

Ready-made cabinet

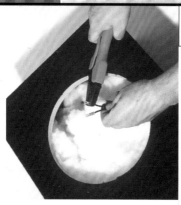

01 The wiring was already connected to the terminal cup that was fitted on the box, so the ends just needed stripping back and terminating with the correct fittings to join onto the subwoofer. The cable is coded with plus and minus symbols for easy connection. As long as you get the feed wire into the box the right way, everything will be fine

02 After carefully positioning the subwoofer to get the centre logo straight, the eight mounting holes were drilled with pilot holes, and then the screws were tightened steadily by hand to stop them getting stripped out. You can use a power screwdriver, but be careful not to go too tight and ruin the pilot hole you've just drilled or you'll have to put your speaker in at a funny angle once you've drilled some more holes

03 The sub cable from the amplifier was clamped tightly under these screw terminals. Like the rest of the wiring, the cable was stripped back and then the ends protected with a short piece of heatshrink tube. This neatens the cable ends and makes it more difficult to short out the wiring. Just be sure to leave enough bare cable to connect to the terminal, eh?

Model specific enclosures

Some sub-box manufacturers offer ready-made vehicle-specific enclosures. These types of enclosures are a convenient alternative for people not willing or able to construct a box, or for those who want the added benefit of a subwoofer but don't want to sacrifice interior space. Most of these enclosures are designed for easy-installation and to blend with the vehicle's interior. They may replace a center console or luggage area side panel - anywhere unused space can turn into sweet sounds.

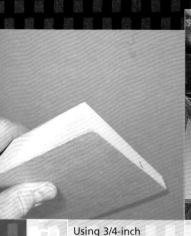   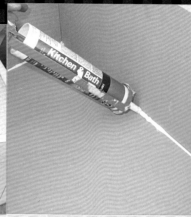

01. Using 3/4-inch Medium Density Fiberboard (MDF) for the enclosure, mark the enclosures measurements and carefully cut the boards

02. Use the template supplied with the subwoofer or the subwoofer mounting ring to mark the positions of the holes, then cut out the holes with a jigsaw

03. When putting the box together, run a bead of glue along the edges of the adjoining seams, then use screws to secure the panels to each other

04. To prevent any air leaks, seal all the seams inside the box with a silicone sealant

Building a sealed enclosure

Warning
Wear a filtering mask before cutting: MDF gives off an extremely fine dust which can be harmful to your health

05. Using spray glue, cover the box with carpet that'll match your interior, then carefully cut out the holes. You'll also have to drill a hole for the speaker wires

06. Follow the manufacturer's instructions for connecting the subwoofer wiring, then place them into the enclosure and mount them, also according to the manufacturer's instructions

07. Be sure to bolt the sub box down so it doesn't roll around. A loose enclosure can be dangerous, particularly in a crash. The last thing you want during an accident is half a ton of unhappy speaker and box come hurtling in the passenger compartment to remind you they weren't bolted down!

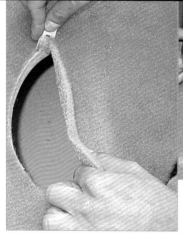  

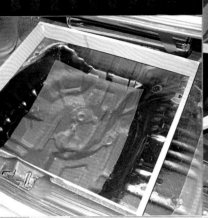   

01 The trunk area was stripped of the spare tire to make room for the enclosure and a support frame was constructed of 3/4-inch Medium Density Fiberboard (MDF)

02 We fabricated rings to hold the subwoofers. The rings are tied in the center by a piece of MDF glued to the bottom, then secured by screws through the top

03 We stretched speaker fabric over the top of the box, secured it with staples, then cut out the holes. Then we brushed a coat of fiberglass resin on both sides

04 After the resin completely dried . . .

Building a custom fiberglass enclosure

 Warning: *Wear a filtering mask before cutting: MDF gives off an extremely fine dust which can be harmful to your health*

05 . . . we reinforced the inside of the box with steel mesh and body filler

06 We also applied a coat of body filler to the exterior to fill in any imperfections . . .

07 . . . then sanded, primered and sanded again for a smooth surface that's ready for the paint shop

08 After returning from the paint shop the box is placed in its final resting spot, then the subs are wired up and mounted

Video

Mobile entertainment

Watch TV. . . pop in a movie on DVD. . . play a video game. . . mobile video now lets you do it all.

The simplest mobile video systems are portable, self-contained units which can be strapped into place between the two front seats, a setup that is popular with minivan owners. There are other systems similar to these, but they are built into the center console and aren't easily removed like the portable type.

The other kind of system is the component type. All component systems begin at the source unit, which could take the form of a remotely mounted VHS tape deck or DVD player, in-dash DVD player, an overhead flip-down screen with an integral DVD player, a Sony PlayStation, a TV tuner or a combination of these. This source signal is then directed to one or more monitors. Some in-dash DVD players have an integral motorized screen that retracts when not in use. When multiple source units are used, they must be connected to a signal distribution box (switcher) installed between the source units and the monitor(s).

Monitor types include the already-mentioned in-dash motorized screen, sunshade monitors, headrest monitors, flip-down overhead console monitors, center console monitors, and monitors that can be mounted on a pedestal or bracket just about anywhere in the vehicle that there's enough room. Just keep in mind that no screen visible to the driver can be operational when the vehicle is in motion.

Video game consoles can be integrated into the system by the use of a signal distribution box and a power inverter that converts 12 volts DC into 110 volts AC. And, with the use of the proper switchbox, a video game can be played on one monitor while a movie is watched on another.

Another neat option that's available with some systems are infrared headphones that allow passengers to listen to the movie or game audio track without the hassle of cords that could get in the way. The infrared signal on these systems is broadcast from transmitters embedded in the monitor housings or from a remote transmitter, usually mounted on the headliner or at the rear of the overhead console where the line-of-sight between the transmitter and headphone will be uninterrupted.

Audio can also be piped through the vehicle's existing speakers. If you've upgraded your audio system with a surround sound system, your passengers will be able to enjoy a near-theatre experience. Just don't let them spill their drinks on the floor, throw bon-bons at the screen or stick their chewing gum to the underside of the seats!

Mobile entertainment

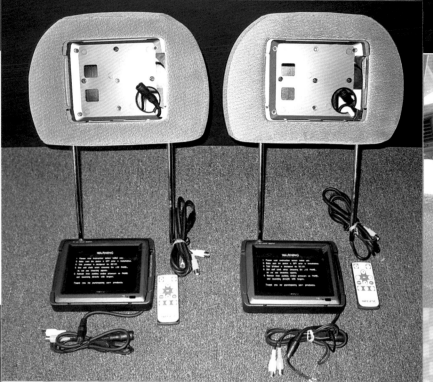

01 So you've decided to install headrest monitors, but cutting into the factory headrests, scooping out some foam, installing a frame then snapping the monitor into the frame is not what you had in mind. The easy option is installing vehicle-specific headrests like these from savv®. The installation requires no disfiguring headrest modification - the vehicle-specific headrests are designed to replace the factory headrests and after a few easy steps a monitor can be installed

02 The cable from the headrest's hollow post was fed down through the seat, then we installed the headrest

Installing a basic video system

Not enough headrests? Install an overhead monitor and keep everyone happy!
Several overhead monitors are supplied with generic housings like this one from savv®. The kit comes with everything needed to install an overhead monitor that looks almost factory **07**

08 A pattern is drawn on the headliner and a hole cut through it for the monitor's mounting bracket and wiring

09 The monitor housing is then attached to the mounting bracket

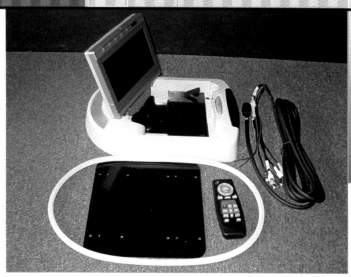

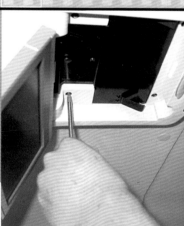

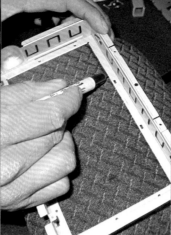

03 Then we placed the monitor into its mounting frame. Wire it up and you're ready to go.

04 If no vehicle-specific headrests with monitors are available for your vehicle, or for some reason you want to retain your headrests, you can modify them to accept video monitors. If you take your time and work carefully you will wind up with a very clean-looking installation, and save a little cash as well. The only real drawback to choosing this route is that the cable to the monitor will be slightly exposed if the headrest is raised

05 Mark the fabric to be cut on the headrest, using the inside of the housing as a template. Be sure to center the housing and make sure it's straight!

06 Connect the A/V cable, then mount the monitor into the housing following the manufacturer's instructions

10 A stand-alone DVD player can be mounted in a convenient location like this under the middle seat of a SUV

11 Follow the manufacturer's instructions for connecting the power wires

12 With the addition of a video selector, one passenger can watch a movie while another plays video games. Of course, some wireless headphones would be in order

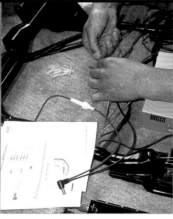

07 Suspension

The main reason for lowering is, of course, to improve your car's looks. Standard suspension nearly always seems to be set too soft and too high - a nicely lowered car really stands out.

Lowering your car should also improve the handling. Dropping the car on its suspension brings the car's center of gravity closer to its roll and pitch centers, which help to hold it to the road in corners and under braking - combined with stiffer springs and shocks, this reduces body roll and increases the tire contact patch on the road. But - if improving the handling is really important to you, choose your new suspension carefully.

If you go the cheap route, or want extreme lowering, then making the car handle better might not be what you achieve.

As for what to buy, there are basically four main options when it comes to lowering:

1 *Set of lowering springs.*

2 *Matched set of lowering springs and shock absorbers.*

3 *Set of "coilovers".*

4 *Air suspension.*

How low to go?

Assuming you want to slam your suspension so that your fenders just clear the tops of your monster new tires, there's another small problem - it takes some inspired guesswork to assess the required drop accurately and avoid the nasty rubbing sound and the smell of burning rubber. Lowering springs and suspension kits will only produce a fixed amount of drop - this can range from 3/4-inch to a more extreme drop of anything up to four inches. Take as many measurements as possible, and ask tuning shops or informed friends. Suppliers and manufacturers are also a good source of help. Coilovers have a range of adjustment possible, which can get you exactly the amount of drop you're looking for; however, this is the most expensive option.

Suspension

High-rate lowering springs

For: This option is inexpensive. You'll get the low-slung look and handling will be slightly improved.

Against: If your shocks are bad or your new springs are badly matched to their damping characteristics, handling could be poor.

Buy: Progressively wound springs which give a smooth ride but also cope with bumps, potholes and extreme cornering; springs for use with standard shocks to prevent pogo-stick handling and shock damage; springs that offer a drop of between 3/4 and 1-1/2 inch - any lower will result in poor handling, fender-rub and excess tire wear.

Matched strut/spring kits

For: Massive improvement in handling thanks to well developed, matched springs and struts. Some struts have adjustable damping so you can fine-tune the ride quality. Price is reasonable.

Against: Due to increased damping and spring rates, the ride may be harsh.

Buy: Kits with progressive springs - these offer an improved ride without compromising handling; kits with adjustable shock damping; kits with multi-position spring platforms - enabling you to tweak the ride height.

Coilover kits

For: Ride height can be adjusted to the level you want. Most kits offer 3/4-inch to 3 inches of adjustment. Damping can be set as desired. The package can be set up to offer awesome handling on the road and then be dumped for shows and cruises.

Against: Coilover packages are expensive and hard to set up properly. If the shocks are set to give a big drop, the springs can pop out of the top cups on full extension. Ride is hard - with a capital H.

Buy: Kits with helper springs - less spring dislocation; kits with adjustable damping.

Air suspension

For: Instant adjustability of ride height; High-Tech look.

Against: Irregular ride and handling; Possible damage to suspension components and body parts if not set up properly.

Buy: Complete kits with compressor, valves, struts and hoses.

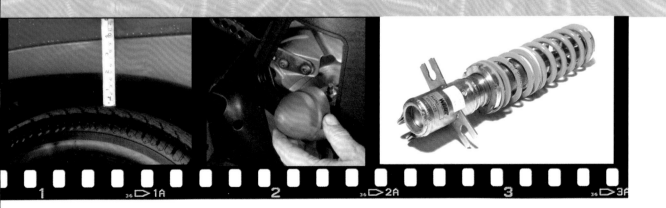

Suspension terms

Fender-rub
Fender-rub occurs when the suspension is too low or too soft. Going over a bump, the wheel is forced up as the suspension compresses, rubbing the wheel opening in the fender.

Circlip
The circlip is a flat, spring-steel clip that fits into a groove on the body of the strut, on which the spring cup sits. On struts designed for lowering, there are sometimes a number of grooves offering differing ride heights.

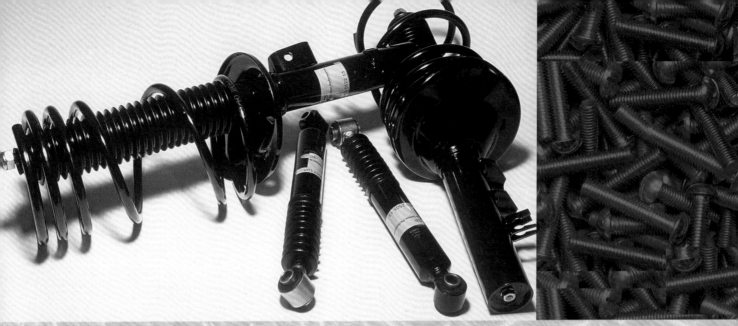

Shock absorbers (shocks)/struts
A shock absorber absorbs the kinetic energy of the spring, damping further reaction from the spring. Shocks, as they're often called, stop the car from bouncing along the road, resulting in better handling. On cars with MacPherson strut-type suspension, the shock absorber and coil spring are incorporated into a strut that is also a structural member of the car's suspension.

Damping
Damping is when the shock absorbs the spring's reaction energy after being compressed. This prevents the spring from releasing its energy rapidly, which would cause a bouncing effect on the car.

Helper spring
Only found on coilover struts, the helper spring prevents dislocation of the main spring when it is unable to cope with extreme extension. The helper spring offers further extension travel, preventing possible dislocation.

Progressive wound
The coils on a progressive spring have been wound closer together at one end. This offers decent ride quality until the loose-wound coils are compressed, then, under further compression, the suspension is stiffer. This limits suspension travel and improves cornering without giving a harsh ride during normal driving.

Ride height
The ride height refers to the height of the chassis from the ground. The height of the springs determines the ride height and center of gravity.

Shot-peened
Shot-peening is a process in which tiny pieces of metal (shot) are fired at the surface of the steel spring, increasing the spring's surface area and making the spring stronger.

Spring cups
The spring cup is located on the body of the strut. The bottom of the spring sits in the cup with the help of an internal lug, which prevents dislocation.

Spring rate
The spring rate translates as the potential resistance against compression of the spring, measured in pounds. A low spring rate means the spring will give a better ride (is "softer"), while a high spring rate means the spring will "give" less and thus has a harsher ride.

Struts and coil springs

Replacing the struts and coil springs is an excellent way to lower a car. Although the car can sometimes be lowered by replacing the springs only, we recommend replacing the strut assemblies with new ones specifically matched to the springs being installed. Doing this will help prevent fender-rub and result in better cornering. When done correctly, this conversion will get you lower but still retain predictable handling characteristics – this is a must if you like to push your car hard through corners.

Many aftermarket companies sell kits of matched struts and springs, often referred to as a "suspension kit." Some of the kits are called "adjustable," but this only applies to the damper rates of the shocks, which can often be customized in a few minutes' work. This feature has no effect on the ride height, but provides the ability to adjust for better ride quality or better handling. We're going to show the work on a car with strut-type suspension. **Note:** *The following is a procedure for replacing a typical front strut/spring. Rear struts and springs disassemble similarly.*

Models with separate strut and knuckle assemblies

01 Loosen the wheel lug nuts, raise the car and support it securely on jackstands. Remove the wheel. Remove the brake hose bracket from the strut (if equipped), as shown. If there is any other wiring that will be in the way of strut removal, such as for the ABS speed sensor, carefully disconnect it and move it out of the way. If the stabilizer bar is attached to the strut, disconnect it, too (see Step 5)

02a If your knuckle is secured to your strut by two bolts, mark around the strut bolt heads for alignment reference on reassembly. Remove the strut-to-knuckle nuts and knock the bolts out with a hammer and punch. Separate the strut from the steering knuckle. Be careful not to overextend the inner CV joint. Also, don't let the steering knuckle fall outward, as the brake hose could be damaged

02b If your knuckle is secured to your strut by a single pinch bolt, remove the bolt (unscrew the nut first, if applicable) and slide the strut body out of the knuckle (you'll have to push down on the knuckle). If it's stuck, you may have to drive a wedge (a large chisel will work) into the split on the backside of the knuckle to free it. Be sure to heed the same precautions as in Step 02a

Models with integral strut and knuckle assemblies

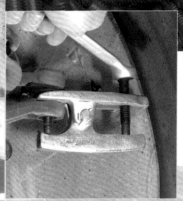

03 On the front wheels of front-wheel-drive cars, loosen the driveaxle hub nuts (there's one in the very center of each wheel). They are on very tight, so use a large breaker bar and be sure the parking brake is securely set. It will often be necessary to unstake the nut with a hammer and chisel (as shown) or remove a cotter pin and nut lock. For clarity, we're showing this with the wheel removed, but it's usually much easier to loosen the nuts with the wheels on the ground. Now raise the vehicle, support it on jackstands and remove the wheels

04 Remove the driveaxle nut. On reassembly, be sure to use a new nut or cotter pin (be sure to stake the nut in the driveaxle groove if it is of that design)

05 If the stabilizer bar is connected to the strut, unscrew the nut and washer and disconnect the stabilizer bar connecting link from the strut body. If the stabilizer bar is connected to the lower arm, unscrew the bolts and washers and remove the mounting clamp from the top of the lower arm

06 Unscrew the tie-rod end nut. Disconnect the tie-rod end using a balljoint separator. Discard the nut and use a new one on installation

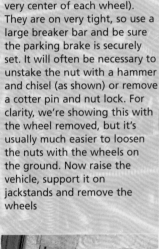

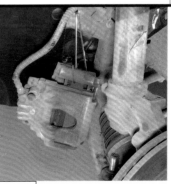

07 Unscrew the nut and remove the lower arm balljoint clamp bolt from the steering knuckle. Discard the nut (a new one should be used on reinstallation)

08 Pry down the lower arm, taking care not to damage the balljoint boot – pry just enough to release the balljoint from the steering knuckle. As you can see we used a chain, block of wood and a metal bar to create a prying device. If the balljoint is a tight fit, open the knuckle clamp a little. Once the balljoint is free, move the strut aside, then release the lower arm slowly

09 Release the hub assembly from the driveaxle by holding the outer CV joint and pulling the strut/knuckle outward. If necessary, the driveaxle can be tapped out of the hub using a hammer and a brass punch. Support the driveaxle by tying it to the car body. Don't allow it to hang down or move outward, as this will damage the inner CV joint or boot. Ensure the brake hose and any wiring is released from the strut/knuckle assembly

10 Loosen and remove the bolts securing the caliper/mounting bracket to the knuckle

11 Slide the brake caliper off the disc, and tie it to the car body to support it. Don't allow it to hang by the flexible hose as this will strain and damage the hose

All models

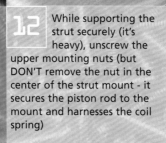

12 While supporting the strut securely (it's heavy), unscrew the upper mounting nuts (but DON'T remove the nut in the center of the strut mount - it secures the piston rod to the mount and harnesses the coil spring)

13 With the nuts removed, lower the strut out from underneath the vehicle. Now, repeat the procedure and remove the other strut

14 Warning: *Disassembling a strut is potentially dangerous and utmost attention must be directed to the job, or serious injury may result.* Before disassembling the struts, make sure you have a good spring compressor and obtain a new piston nut for each strut (the nut should be replaced every time it is removed). Install the spring compressor according to the manufacturer's instructions and compress the coil spring until all spring pressure is relieved from the seats. Ensure the spring compressor is securely installed and in no danger of slipping off before proceeding

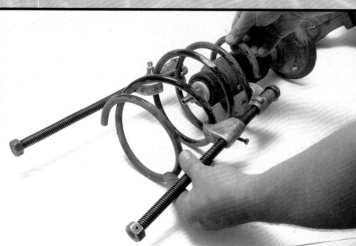

17 ... followed by the bearing top plate, then the bearing ...

18 ... followed by the spring seat

19 Remove the coil spring, complete with spring compressor. After removing the coil spring from the strut assembly, set it aside in a safe, isolated area

15 Remove the trim cap, then loosen and remove the piston nut and its collar. If necessary, retain the piston with a Torx or Allen wrench to prevent rotation while loosening the nut. Discard the nut (a new one should be installed)

16 Lift off the strut upper mounting plate . . .

20 Slide the rubber boot off the strut, along with the rubber bumper and its collar. Note how all the parts are installed so you can assemble them the same way

21 Assemble the strut by reversing the disassembly sequence. Be sure the piston nut is tightened to specification and the spring ends are properly seated in the spring seat, as shown. Then slowly release the spring compressor. The remainder of assembly is the reverse of disassembly. Be sure to tighten all fasteners to specification. Have the wheel alignment checked and if necessary, adjusted

Rear coil spring Suspension

Although most sport-compact cars use strut-type suspension front and rear, some use separate coil springs and shock absorber assemblies for the rear suspension. Replacing the springs and shocks on these models is relatively simple.

Tricks 'n' tips
A coil spring compressor usually isn't needed, but you will need to support the rear suspension arms. Ideally, you should have two floor jacks and an assistant, as both sides usually have to be lowered at the same time. There's a risk of snapping the brake lines if the rear suspension's just allowed to drop down unsupported. Without support, the rear arms will drop as soon as the shock absorber mounts are removed.

01 Loosen all the rear wheel lug nuts, jack up each rear corner of the car and support it securely with jackstands under each rear jacking point. Remove both rear wheels. Using a jack directly under each suspension arm, raise the rear arms slightly, just enough to take the load off the shock mounts. Now find the upper shock mounting points (usually in the trunk or rear cargo area, sometimes behind a trim panel or cover)

02 If equipped, pull off the rubber cap installed over the top nut

03 Using two wrenches, loosen and remove the shock top locknut and the main nut underneath

04 With the nuts removed, take off the top plate/washer and the rubber bushing below. Repeat this process on the other side

05 Back down below, and making sure that the suspension arm is supported, loosen and remove the shock lower mounting bolt. If it's stubborn, apply some penetrating oil and let it soak in, then try again. These bolts are prone to rust and are sometimes difficult to remove

06 To remove the rear shock, it will be necessary to lower the suspension arm slightly until it's free. You shouldn't have to lower the arm very far for this, but if you do, keep an eye on the brake lines, making sure they're not placed under any strain. Pull the shock down to free it from the top mounting, and remove it from the vehicle. Repeat the process on the other side, so that both shocks are off. On some models, the suspension arms are joined along a common shaft, so you'll have to lower the jacks simultaneously – it's best to have an assistant in this case

07 To avoid damaging the rear brake lines as the rear axle is lowered, they must be disconnected from the suspension arms. Here, a horshoe clip secures the brake hose to the arm. Also look for wiring or other hardware that will need to be disconnected from the arm before it's lowered

08 If the brake lines travel down the suspension arm, unclip them so they won't be bent when the suspension arm is lowered

09 Slowly and carefully lower the jacks under the suspension arms. Stop every inch or so, and check that the brake lines don't get damaged. Eventually, the rear springs can be removed by hand

10 When installing the new springs, pay attention to the markings that show which end goes up. Install the springs to the molded-in lower spring seat . . .

11 . . . then carefully raise both jacks, watching the brake lines as before. Make sure the tops of the springs engage properly with the spring seats

12 If you're installing adjustable shock absorbers, pre-set the adjustment to the basic setting recommended by the shock manufacturer.

13 Install the bottom end of the shock into place on the suspension arm, and insert the bolt.

14 Don't tighten the lower shock bolts completely at this point – wait until the shock is completely installed and the jacks are removed, then tighten to specification. Or, if access under the vehicle is difficult, you can raise the suspension with the floor jack to simulate normal ride height, then fully tighten the bolts (after steps 16 and 17 have been performed)

15 Don't forget to reattach the brake lines to the clips

16 Raising the suspension arm as necessary, feed the shock absorber up into its mounting hole

17 Back in the trunk again, install the rubber bushing, top plate/washer, and finally, the main nut and locknut. Tighten the main nut securely, then tighten the locknut against it, using two wrenches. The remainder of installation is the reverse of removal. Be sure to tighten all fasteners securely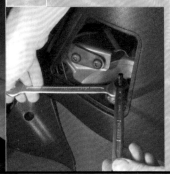

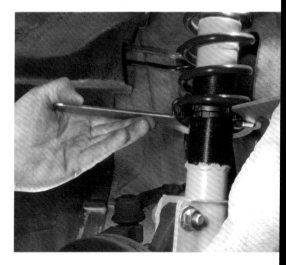

Coilovers

If you've chosen coilovers, you obviously know quality when you see it, and you're not prepared to compromise. True, quality costs, but you get what you pay for. This is an expensive option, but it offers one vital feature that others can't - true adjustability of ride height. This means you can get exactly the ride height you're looking for. Coilovers are a variation on the suspension kit theme, in that they are a set of matched springs and struts, but with the added bonus of being fully adjustable (within certain limits, obviously).

Coilover conversion

Another option gaining popularity is the "coilover conversion." If you must have the lowest, baddest machine and want to save some money, these could be the answer. Offering as much potential for lowering as genuine coilovers, these items could be described as a cross between coilovers and lowering springs - the standard struts are retained (which usually results in less ride quality). What you get is a new spring assembly, with adjustable top and bottom mounts - the whole thing slips over your standard strut. Two problems with this solution:

- Standard struts are not designed to operate as well when lowered, so the car's ride and handling will be compromised if you lower the car very much.
- The standard struts are effectively being compressed, the lower you go. There is a limit to how far they will compress before being completely solid. Needless to say, even a partly-compressed strut won't be able to do much actual damping - the result could be a very harsh ride.

Air suspension

Air suspension systems provide the ability to raise or lower the car instantly, from the driver's seat, or even from a distance on systems equipped with a remote control. Many consider this the ultimate suspension system, since you can lower the car as low as you want for shows and cruises, but still be able to easily raise the car to a reasonable ride height for your daily commute. The main components are air "springs" (which are basically strut assemblies with no coil springs) filled with compressed air. Pumping air into the air springs increases the air pressure and raises the vehicle, while releasing air from the springs lowers the vehicle. The system also has an electric air compressor, valves and air lines to deliver and release the compressed air. Air suspension systems also have a high-tech look: the compressors are often highly polished and sometimes chrome or gold plated – they can really dress up your engine compartment!

The primary disadvantage of an air suspension system, as you might have guessed, is cost: generally, they have a similar cost to coilover kits. And the ride quality and handling will never reach the level of a good, well-adjusted coilover kit – in fact, ride quality will likely be worse than with the stock suspension system. Another disadvantage is that the systems can cause damage to your vehicle's suspension and body if they are not set up and used carefully. The systems should be set up so that there will be no fender rub or metal-to-metal suspension bottoming when the air is released. Sometimes systems can leak down when the vehicle is left sitting for an extended period - you don't want to wind up with bent parts as a result.

There are a variety of air suspension systems available, and installation procedures vary. The air "springs" install basically the same way as standard struts or shocks. The air compressor is generally installed under the hood. Many owners choose to install a large separate storage tank for compressed air that allows rapid, repeated cycling of the system ("bouncing" the car up and down). When installing this equipment, refer to the manufacturer's instructions and recommendations.

Nasty side-effects

Camber angle and tracking

With any lowering "solution," your suspension and steering geometry will be affected - this will be more of a problem the lower you go. This will manifest itself as steering that either becomes lighter or (more usually) heavier, and as tires that wear out quickly on the inner or outer edges. If you've spent a small fortune on your wheels and tires, as most of us do, think carefully about this tire wear issue. Whenever you've installed a set of springs (and this applies to all types), have the wheel alignment checked ASAP afterwards.

Rear brake proportioning valve

Some cars have a Load Sensing Proportioning Valve (LSPV) installed in the rear brake line, linked to the rear suspension. When the car is lightly loaded, the braking effort to the rear is limited, to prevent the wheels locking up. With the trunk full of luggage, the back end sinks down, and the valve lets full braking pressure through to the rear. When you slam the suspension, the valve is fooled into thinking the car is always loaded, and you might find the rear brakes locking up unexpectedly - could be a nasty surprise on a wet road.

Some owners of lowered cars disconnect this valve from the suspension and try to find a "central" position they can lock the valve into. The theory is that they have now simulated the normal "unloaded" position of the valve and will now have normal braking. While this might work in some cases, the safest approach would be to have a factory-authorized shop properly set the valve for the amount of lowering you've done.

Suspension

Strut brace

When you pitch your car into a corner hard, and then start sawing the wheel back and forth as you negotiate a tricky series of left and right hand corners, the loads imposed on the front end is considerable enough to actually twist the body, which will change the relationship of the struts or upper control arms to each other. You don't want that!

If you've installed an aftermarket stabilizer bar, this problem will actually be *magnified*, because of the greater cornering forces you'll be imposing on the chassis. The camber and caster of the front wheels is established by the position of the upper ends of the struts. Allowing the upper mounting points of the struts to move means that the camber and caster are changing. A strut brace simply ties the two strut towers together and prevents them from flexing. And that's a good thing!

01 This Neuspeed kit uses aircraft grade bolts and "nutserts" to screw them into. Drilling holes for the bolts, then installing a couple of conventional nuts on them from underneath is difficult on any model and impossible on some. Nutserts eliminate that problem; you simply drill a hole for each nutsert, install the nutsert and then use it as the "nut" for each bolt

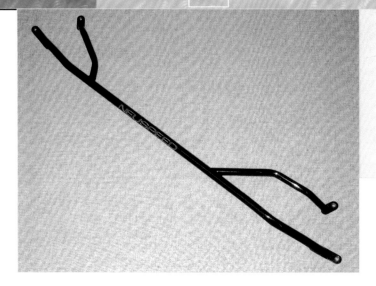

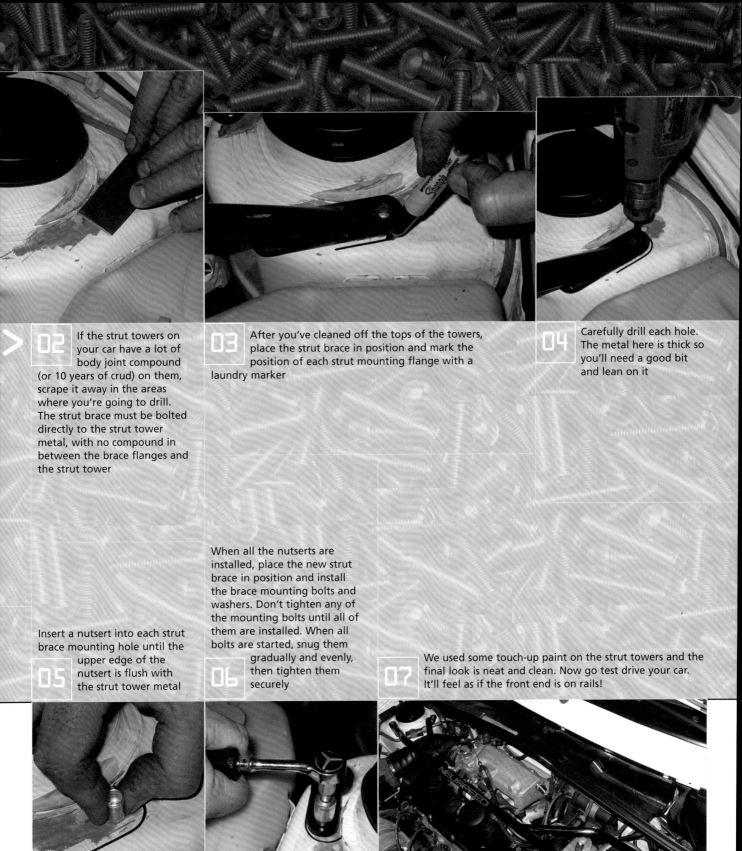

02 If the strut towers on your car have a lot of body joint compound (or 10 years of crud) on them, scrape it away in the areas where you're going to drill. The strut brace must be bolted directly to the strut tower metal, with no compound in between the brace flanges and the strut tower

03 After you've cleaned off the tops of the towers, place the strut brace in position and mark the position of each strut mounting flange with a laundry marker

04 Carefully drill each hole. The metal here is thick so you'll need a good bit and lean on it

05 Insert a nutsert into each strut brace mounting hole until the upper edge of the nutsert is flush with the strut tower metal

06 When all the nutserts are installed, place the new strut brace in position and install the brace mounting bolts and washers. Don't tighten any of the mounting bolts until all of them are installed. When all bolts are started, snug them gradually and evenly, then tighten them securely

07 We used some touch-up paint on the strut towers and the final look is neat and clean. Now go test drive your car. It'll feel as if the front end is on rails!

Suspension

Installing stabilizer bars

Even with stock suspension, most sport compact cars are fun to fling around. But their handling limitations become apparent when you start flicking them into corners a little harder than usual.

Stabilizer bars reduce the tendency of the car's body to "roll" (or tilt towards the outside of a turn) when cornering. Your vehicle probably already has a stabilizer bar at the front, and maybe even one at the rear. But even if you have stabilizer bars front and rear from the factory, a good set of aftermarket performance bars will do you a world of good - you won't believe how flat your car will corner!

We're showing a typical installation, but each car is a little different. A Haynes manual for your make and model will give you detailed step-by-step instructions for your particular car.

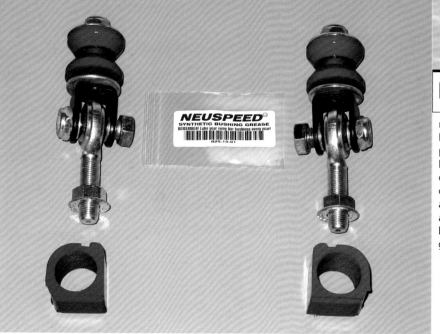

01 Unpack your stabilizer bar kit and make sure that everything you'll need is there. This Neuspeed kit includes a pair of polyurethane bar clamp bushings and a pair of links (to connect the stabilizer to the control arms) with rod ends and polyurethane bushings. It also includes a couple of small bags of a special synthetic grease

02 Raise the vehicle and place it securely on jackstands. Locate the stock link at each end of the stabilizer bar, note the location of the rubber bushings and spacers (you're going to install the new bushings in the exact same order)

03 Remove the nut from the bottom and disconnect the link from the control arm, and if you can, remove the link from the end of the stabilizer. If it's difficult to remove the old links right now, don't worry about it, just push the links out of the way so that they don't interfere with removal of the old stabilizer bar

04 On a Golf (and on many other vehicles), the suspension pieces (including the front stabilizer) are hung from or attached to a subframe. On this Mark II Golf, you'll need to lower the front subframe enough to remove the old stabilizer bar and install the new unit

Understeer and oversteer

What exactly is "understeer"? And what is "oversteer"? A good understanding of oversteer and understeer is essential to anyone planning to upgrade the tires, wheels, coil springs and front and rear stabilizers.

Understeer

Understeer is the tendency of a car to slide off the road nose end first when you pitch it into a corner too hard. You turn the steering wheel as you enter the corner, but the car turns less than it should in response to the amount of steering input you're giving it. Most modern sport compact cars, particularly front-wheel-drive vehicles, are set up by the manufacturer for some understeer because at normal cornering speeds a bit of understeer feels safer and more predictable. But if you want to go fast through corners, and not just in a straight line, you will have to eliminate your car's understeer tendencies. One way to do that is to install a rear stabilizer bar or, if your car is already equipped with one, a bigger rear bar.

Oversteer

Oversteer is the tendency of a car to slide off the road tail end first when you pitch it into a corner too hard. You turn the steering wheel as you enter the corner, but the car turns more than it should in response to the amount of steering input you're giving it. An oversteering car can be quick through the corners, but it can also be dangerous unless it's in the hands of a very skilled driver. Luckily, very few front-wheel-drive sport compact cars have an oversteer problem, because their weight distribution is invariably biased in favor of the front end. Getting a stock front-wheel-drive sport compact car to oversteer requires that you get into a corner way too hot, then abruptly lift your foot off the accelerator pedal.

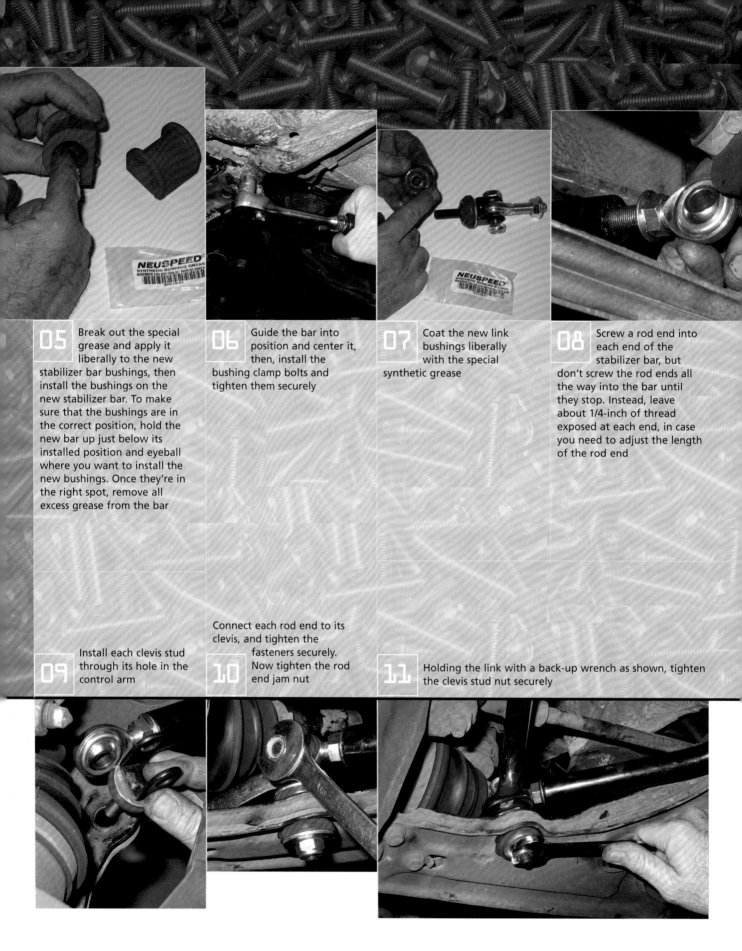

05 Break out the special grease and apply it liberally to the new stabilizer bar bushings, then install the bushings on the new stabilizer bar. To make sure that the bushings are in the correct position, hold the new bar up just below its installed position and eyeball where you want to install the new bushings. Once they're in the right spot, remove all excess grease from the bar

06 Guide the bar into position and center it, then, install the bushing clamp bolts and tighten them securely

07 Coat the new link bushings liberally with the special synthetic grease

08 Screw a rod end into each end of the stabilizer bar, but don't screw the rod ends all the way into the bar until they stop. Instead, leave about 1/4-inch of thread exposed at each end, in case you need to adjust the length of the rod end

09 Install each clevis stud through its hole in the control arm

10 Connect each rod end to its clevis, and tighten the fasteners securely. Now tighten the rod end jam nut

11 Holding the link with a back-up wrench as shown, tighten the clevis stud nut securely

12 The rear stabilizer bar kit includes a pair of polyurethane bar clamp bushings and a pair of links (to connect the stabilizer to the rear axle trailing arms) with rod ends. It also includes a couple of small bags of a special synthetic grease

13 Coat the new bushings with the special synthetic grease and install them on the bar.

14 With the bushings correctly positioned on the stabilizer bar, lift the bar back into place and install the bushing clamps

On this Golf we had to install the bolt through the top of each clamp, then screw it into the small plate (included in the kit for each clamp). This step was tricky because these new bushings are hard! They don't compress easily, which is what you want, but they're difficult to squeeze tightly enough to get the upper tip of the clamp over the top of the axle beam and far forward enough to get the bolt and plate on. You'll definitely want a helper for

15 this step

Install the link as shown, and tighten the bolt and nut securely. (On the Neuspeed setup, the middle hole of the bar is the recommended

16 starting point)

On this installation it was necessary to remove the lower shock absorber mounting nut and bolt (just do one side at a time - don't remove the other lower shock nut and bolt until you've finished

17 this side) . . .

. . . because a longer bolt was required (it also secures the top of the link). Do the other side and

18 your done!

08 Wheels & tires

Nice custom wheels with wide, low-profile tires are the number one gift you can give to your car. As a bonus, big wheels with the right tires can actually provide a huge improvement in your car's handling. But before you slap down that serious cash, figure out what will look good and work well on your car.

If you want the ultimate look, you'll get the tires and wheels that are as close as possible to looking like a rubber band has been wrapped around a 21-inch wheel. Not only is this nice to look at, it's race-inspired. In hard cornering, stock tires do not stay centered on the wheel. They "roll under," so that the tread area actually moves away from the center of the wheel. When this happens, the tread distorts, the tire "breaks away" and starts sliding. On low-profile tires, roll-under is almost non-existent, so the tire can keep a consistent tread "patch" on the road. Additionally, the treads of low-profile tires are often designed for maximum adhesion, so they stick to the road better. Tires designed for long tread life are usually made from a very hard rubber compound and therefore will last a very long time. But hard-compound tires don't stick to the road as well. So most high-performance low-profile tires are made from softer, stickier rubber compounds that will allow better traction for cornering and acceleration - but the tires will tend to wear out faster.

As you can see, tire design is compromise. No one tire design can do everything well. Think about how much you really want to spend on your tires and wheels, including how often you can afford to replace the tires. And think about ride quality, too. You may not be an old man with a bad back, but if you do a lot of driving, constant bumping and bouncing can be annoying. Remember that the ultra-low profile tires are not very practical if you do a lot of driving on rough roads. One good pothole hit is all it takes to bend the rim on a bucks-up wheel. A cheaper, higher-profile tire and wheel might absorb a hit from a pothole or curb just fine.

So don't just go into a showroom and point to the wheels and tires that look the best – with that level of research, you're almost sure to be disappointed. Talk to the salespeople and technicians at the tire store and find out about the ride quality and treadlife of each tire, as well as its cost. Tire technicians who install a lot of tires can help you figure out how big a tire/wheel combination you can fit in your car. And if you're planning on lowering the car, tell them about that, too, since it will have an effect on maximum tire size. When in doubt about maximum tire size, contact the tire manufacturer for their recommendations.

Many people try to measure their own wheelwells with a tape-rule to determine how much clearance they have for larger tires and wheels. Even if you take these measurements with the front wheels in their extreme turn positions, there are many other variables that are very hard to calculate. For example, the new wheels will likely have more offset than your existing wheels. Overall tire-and-wheel height and width will be different than the wheel measurements themselves. And don't forget about suspension movement – you don't want to have the tires rub everytime you go over a bump! No, it's best to let the experts determine your maximum tire and wheel size, and they are usually happy to help. And if the combination doesn't work, you'll have someone to blame beside yourself!

Offset

Simply put, offset is the distance from the centerline of the wheel to the surface of its mounting face. Offset permits the use of wide tires and wheels without extending the wheels outward, past the fender area. Basically, you're setting the wheel back, into the wheelwell instead of outward, where you'll poke the wheel out past the fender, which not only looks funny, but is illegal in many areas. By using up your wheel-well space with offset wheels, you can get your wide tires onto your car without rubbing them on the fender lips. While this sounds simple, selecting the offset that will allow you to have the widest wheel possible gets a bit complicated and you need to spend some time figuring out the specifics for your vehicle.

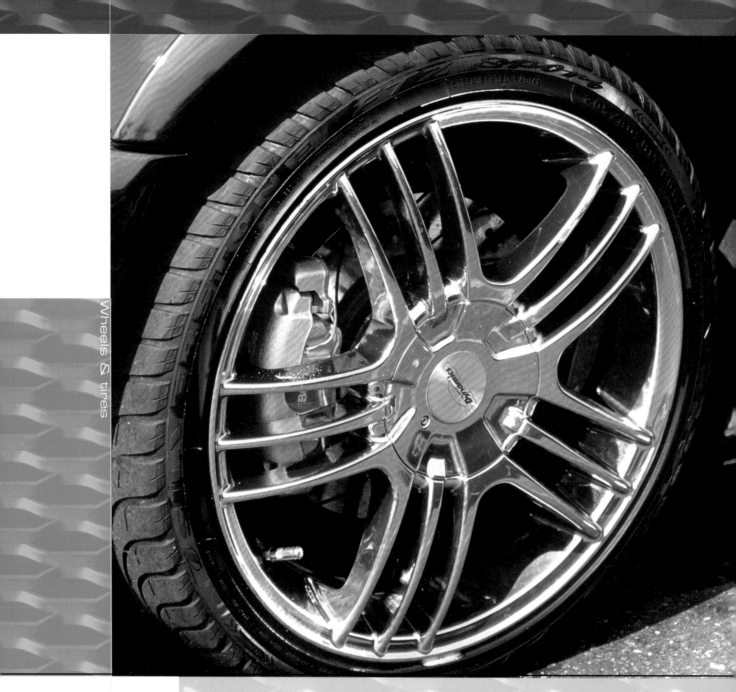

Rolling radius

Rolling radius is the distance from the center of the wheel to the edge of the tire's tread. Increasing it increases the time taken for the wheel to rotate one turn.

Installing bigger wheels on your car can critically affect its gearing unless you choose tires to match. This means your speedometer won't read correctly, and you could lose acceleration.

To solve any problems, choose lower-profile tires to go with larger-diameter wheels. If you go up one rim size, decrease profile one size.

Confused? Well, a 205/55x15 tire (that's a 15-inch wheel, obviously) has a rolling radius of 607mm. Upgrading to a 17-inch wheel means fitting a 215/40x17 rubber with a nearly-identical rolling radius of 606mm. Stay within ten percent and you should be okay.

Tire size markings

All tires carry standard tire size markings on their sidewalls, such as **"195/60 R 15 87H"**.

195   indicates the width of the tire in mm.

60   indicates the ratio of the tire section height to width, expressed as a percentage. If no number is present at this point, the ratio is considered to be 82%.

R   indicates the tire is of radial ply construction.

15   indicates the wheel diameter for the tire is 15 inches.

87   is an index number which indicates the maximum load that the tire can carry at maximum speed.

H   represents the maximum speed for the tire which should be equal to or greater than the car's maximum speed.

Note that some tires have the speed rating symbol located between the tire width and the wheel diameter, attached to the "R" radial tire reference, for example, "195/60 HR 15".

Speed rating symbols for radial tires

Symbol	mph
P	93
Q	99
R	106
S	112
T	118
U	124
V (after size markings)	Up to 150
H (within size markings)	Up to 130
V (within size markings)	Over 130
Z (within size markings)	Over 150

Bolt pattern

Another thing to worry about when selecting wheels is the bolt pattern. Most cars have four or five nuts or bolts holding each wheel to its hub. Different manufacturers use different bolt patterns for their wheels, so, just because two wheels have the same number of lug nuts, does not mean they will interchange. Just another complication. So before you buy used wheels from your buddy, be sure they will fit.

Gallery of Wheels

Wheels & Tires

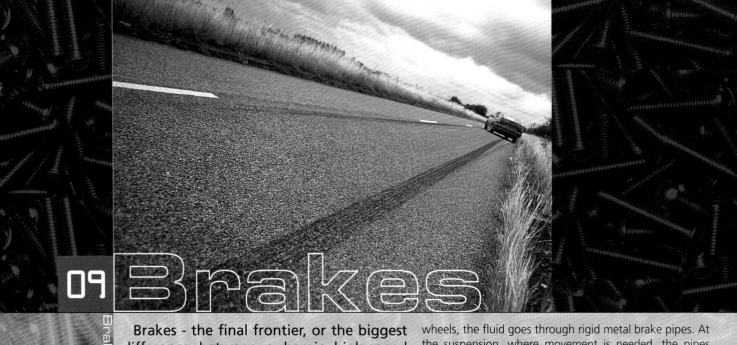

09 Brakes

Brakes - the final frontier, or the biggest difference between a heroic high-speed charge and a humiliating high-speed crash. Here's what's what...

Brake pedal
Where it all starts, from your point of view. The pedal itself is a mechanical lever, and has to provide enough leverage to work the brakes if the power booster fails. Manufacturers use "brake pedal ratios" to express this - a low-ratio pedal, for instance, will give quick-acting but hard-to-work brakes.

Booster
The power brake booster on most cars is vacuum-operated, and it produces extra force on the piston inside the master cylinder when the brakes are applied, reducing brake pedal effort. Most cars these days are equipped with boosters because they are also equipped with disc brakes (at least up front), and disc brakes require more force than drum brakes do (they aren't self-energizing).

Master cylinder
Below the brake fluid reservoir is the master cylinder, which is where the brake pedal effort (force) is converted to hydraulic effort (pressure), and transmitted to each brake through the hydraulic lines and hoses.

Brake fluid
Pressing the brake pedal moves the brake fluid through the lines to each brake, where the hydraulic pressure is converted back to mechanical force once more, as the pads and shoes move into contact with discs and drums. Fluid does not compress readily, but if air (which is compressible) gets in the system, there'll be less effort at the brakes and you'll get a fright. This is why 'bleeding' the brakes of air bubbles whenever the system has been opened is vital.

Brake hoses
For most of the way from the master cylinder to the wheels, the fluid goes through rigid metal brake pipes. At the suspension, where movement is needed, the pipes connect to flexible hoses. The standard rubber hoses are fine when they're new, but replacing old ones with great-looking braided hoses is a good move - in theory, it improves braking, as braided hoses expand less than rubber ones, and transmit more fluid pressure.

Calipers
These act like a clamp to force the brake pads against the discs - fluid pressure forces a piston outwards, which presses the brake pads onto the disc. Generally, the more pistons your calipers have, the more surface area the brake fluid has on which to push, providing greater clamping power. Most cars have only one or two per caliper, while exotic calipers may contain three, four or even more!

Discs
Clamped to your wheel hubs, the discs spin around as fast as your wheels. Over 90% of braking is done by the front brakes, which is why all modern cars have front discs - they work much better than drums, because they dissipate heat better. Many base models have "solid"

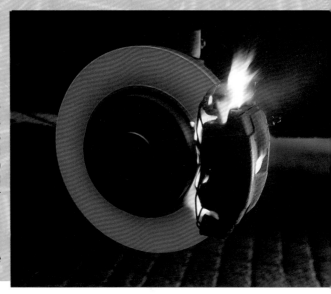

discs, while higher-performance cars have "vented" ones, which have an air gap between the braking surfaces to aid cooling. The bigger the disc diameter, the greater pad area available, with greater stopping power (think of a long lever as compared to a short one).

Pads and shoes

Both have a metal backing plate, with friction lining attached. Brake linings used to be made of an asbestos compound, which had excellent heat-resisting qualities but could cause cancer or asbestosis in people who breathed in the dust. Now they come in non-asbestos organic (stock), or for higher-performance applications, semi-metallic or carbon metallic.

Anti-lock Brakes (ABS)

Bottom line: Two cars are cruising down the road and a hay truck pulls out from a side road. Both drivers stand on the brake pedal. The car without ABS "locks up" the wheels and starts skidding, ending up sideways in a ditch. The car with ABS is able to slow down and safely maneuver around the truck. Great stuff, and here's how it works.

ABS works by detecting when a particular wheel is about to lock. It then reduces the hydraulic pressure applied to that wheel's brake, releasing it just before the wheel locks, and then re-applies it.

The system consists of a hydraulic unit, which contains various solenoid valves and an electric fluid return pump, four roadwheel sensors, and an electronic control unit (ECU). The solenoids in the hydraulic unit are controlled by the ECU, which receives signals from the four wheel sensors.

If the ECU senses that a wheel is about to lock, it operates the relevant solenoid valve in the hydraulic unit, which isolates that brake from the master cylinder. If the wheel sensor detects that the wheel is still about to lock, the ECU switches on the fluid return pump in the hydraulic unit and pumps the fluid back from the brake to the master cylinder, releasing the brake. Once the speed of the wheel returns to normal, the return pump stops and the solenoid valve opens, allowing fluid pressure back to the brake, and so the brake is re-applied. Pretty impressive, especially when you consider all this is happening in a fraction of a second! You

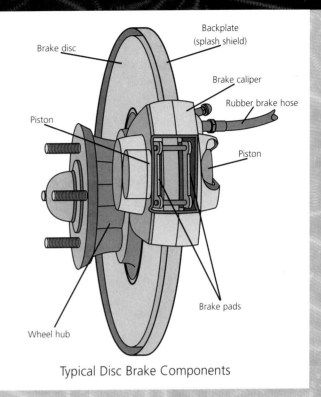

Typical Disc Brake Components

may feel your ABS system working in a hard braking situation. The brake pedal will "pulse" as the pressure varies. But don't let up or "pump" the pedal yourself - let the ABS do its job. The rapid variations in fluid pressure cause pulses in the hydraulic circuit, and these can be felt through the brake pedal.

The system relies totally on electrical signals. If an inaccurate signal or a battery problem is detected, the ABS is automatically shut down, and a warning light on the instrument panel will come on. Normal braking will always be available whether or not the ABS is working.

ABS cannot work miracles, and the basic laws of physics will still apply: stopping distances will always be greater on slippery surfaces. The greatest benefit of ABS is being able to brake hard in an emergency without having to worry about correcting a skid.

If you have any problems with an ABS, always consult an authorized dealer.

Upgraded discs and pads are often the first step to a high-performance brake system, a simple modification that takes a couple of hours max, and can make a huge difference. Since they do wear down over time, and are also prone to warpage, there's a chance you'll have to change your discs anyway, so why not upgrade them at the same time?

Upgraded discs and pads

01 Loosen the wheel lug nuts or bolts, raise the front of the vehicle and support it securely on jackstands. Pry off the anti-rattle spring if your caliper is equipped with one.

02 Unscrew the two caliper guide bolts.

03 Lift off the caliper and the inner brake pad. . .

Some good things to know about working on brakes

- Brakes create a lot of dust from the friction linings. Although usually not made from asbestos (very bad stuff) anymore, the dust is still something you'll want to avoid. Spray everything with brake cleaner and don't blow it into the air where you'll breathe it.
- Brake fluid is nasty stuff - poisonous, highly flammable and an effective paint stripper. Mop up spills promptly and wash any splashes off paintwork with lots of water.
- Do not use petroleum-based cleaners and solvents on or around brake parts. It will eat away all the rubber parts and hoses. Use only brake cleaner.
- Most brake jobs you can do without loosening or removing the fluid hoses and lines. If you mess with these you'll let air into the system and then have to "bleed" the brakes, which can be tricky. If you finish your job and then step on the pedal and it goes all the way to the floor, or feels soft or "spongy," you've let air into the lines.
- Which brings up a good point. After working on your brakes, start the car and pump the pedal a few times to make sure all is well before charging down the street.
- Finally, what you see in this book should be considered "typical." All cars are different, which is why you should have a Haynes Automotive Repair Manual for your specific vehicle.

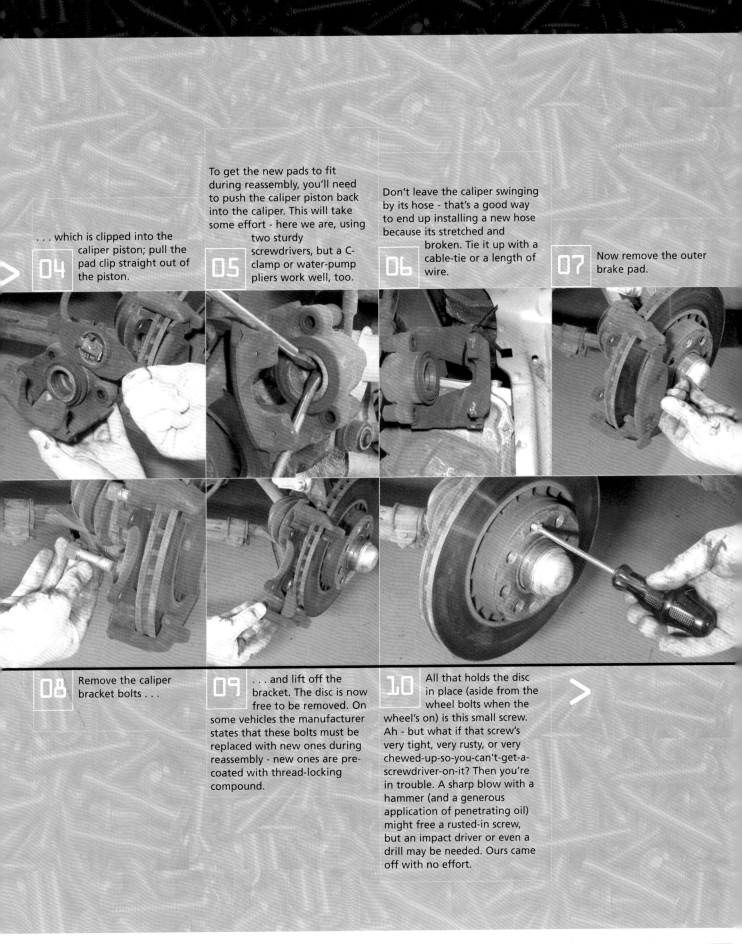

04 ...which is clipped into the caliper piston; pull the pad clip straight out of the piston.

05 To get the new pads to fit during reassembly, you'll need to push the caliper piston back into the caliper. This will take some effort - here we are, using two sturdy screwdrivers, but a C-clamp or water-pump pliers work well, too.

06 Don't leave the caliper swinging by its hose - that's a good way to end up installing a new hose because its stretched and broken. Tie it up with a cable-tie or a length of wire.

07 Now remove the outer brake pad.

08 Remove the caliper bracket bolts...

09 ...and lift off the bracket. The disc is now free to be removed. On some vehicles the manufacturer states that these bolts must be replaced with new ones during reassembly - new ones are pre-coated with thread-locking compound.

10 All that holds the disc in place (aside from the wheel bolts when the wheel's on) is this small screw. Ah - but what if that screw's very tight, very rusty, or very chewed-up-so-you-can't-get-a-screwdriver-on-it? Then you're in trouble. A sharp blow with a hammer (and a generous application of penetrating oil) might free a rusted-in screw, but an impact driver or even a drill may be needed. Ours came off with no effort.

149

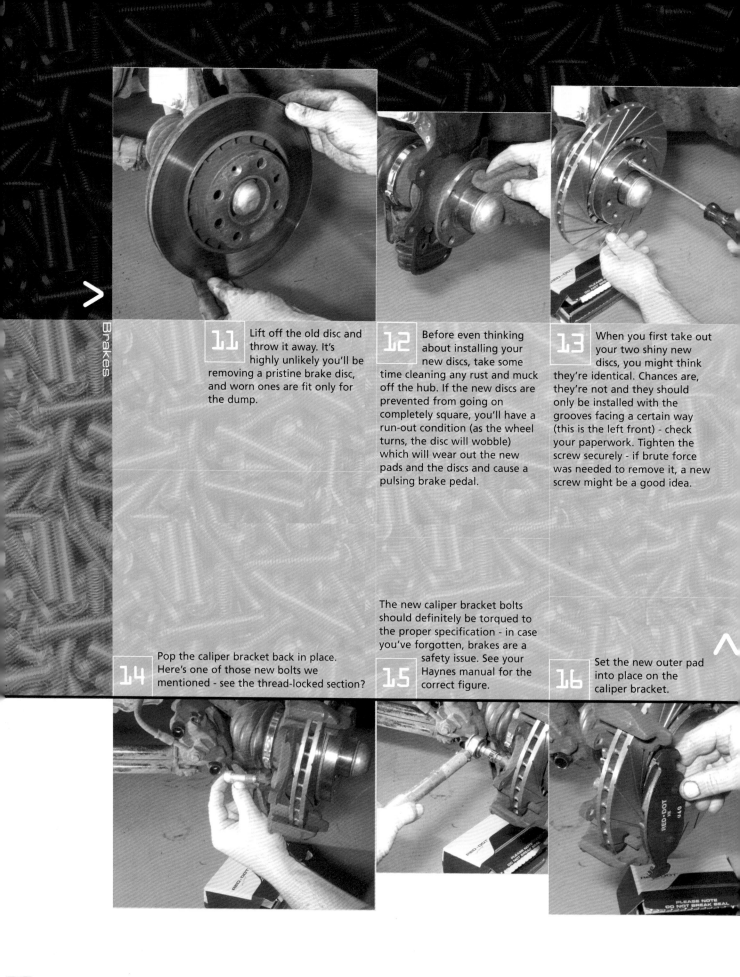

11 Lift off the old disc and throw it away. It's highly unlikely you'll be removing a pristine brake disc, and worn ones are fit only for the dump.

12 Before even thinking about installing your new discs, take some time cleaning any rust and muck off the hub. If the new discs are prevented from going on completely square, you'll have a run-out condition (as the wheel turns, the disc will wobble) which will wear out the new pads and the discs and cause a pulsing brake pedal.

13 When you first take out your two shiny new discs, you might think they're identical. Chances are, they're not and they should only be installed with the grooves facing a certain way (this is the left front) - check your paperwork. Tighten the screw securely - if brute force was needed to remove it, a new screw might be a good idea.

14 Pop the caliper bracket back in place. Here's one of those new bolts we mentioned - see the thread-locked section?

15 The new caliper bracket bolts should definitely be torqued to the proper specification - in case you've forgotten, brakes are a safety issue. See your Haynes manual for the correct figure.

16 Set the new outer pad into place on the caliper bracket.

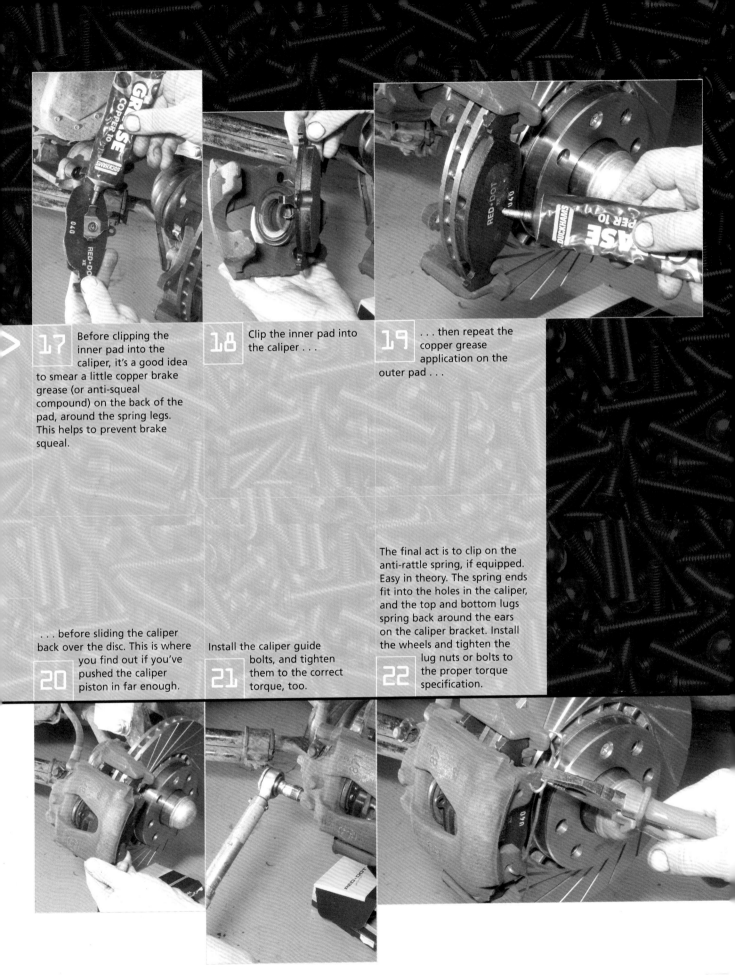

17 Before clipping the inner pad into the caliper, it's a good idea to smear a little copper brake grease (or anti-squeal compound) on the back of the pad, around the spring legs. This helps to prevent brake squeal.

18 Clip the inner pad into the caliper . . .

19 . . . then repeat the copper grease application on the outer pad . . .

20 . . . before sliding the caliper back over the disc. This is where you find out if you've pushed the caliper piston in far enough.

21 Install the caliper guide bolts, and tighten them to the correct torque, too.

22 The final act is to clip on the anti-rattle spring, if equipped. Easy in theory. The spring ends fit into the holes in the caliper, and the top and bottom lugs spring back around the ears on the caliper bracket. Install the wheels and tighten the lug nuts or bolts to the proper torque specification.

Brakes

Bigger brakes

Along the same lines as installing upgraded discs and pads is the addition of bigger brakes. A larger diameter brake is a more powerful brake. Think of it as a lever - a longer lever will enable you to move a heavier object than a short one would, right? Well, a bigger brake will soak up more torque and slow the vehicle down more quickly. It'll also operate at lower temperatures than a standard diameter brake.

01 Loosen the caliper mounting bolts (but don't remove them yet). On this setup we had to hold the caliper slide pin still with one wrench while loosening the mounting bolt with another. The bolt was also "frozen," so we had to give it a sharp rap with a hammer.

05 If you have a hose clamp like this, screw it down onto the brake hose. If you don't, cap the end of the line after detaching it.

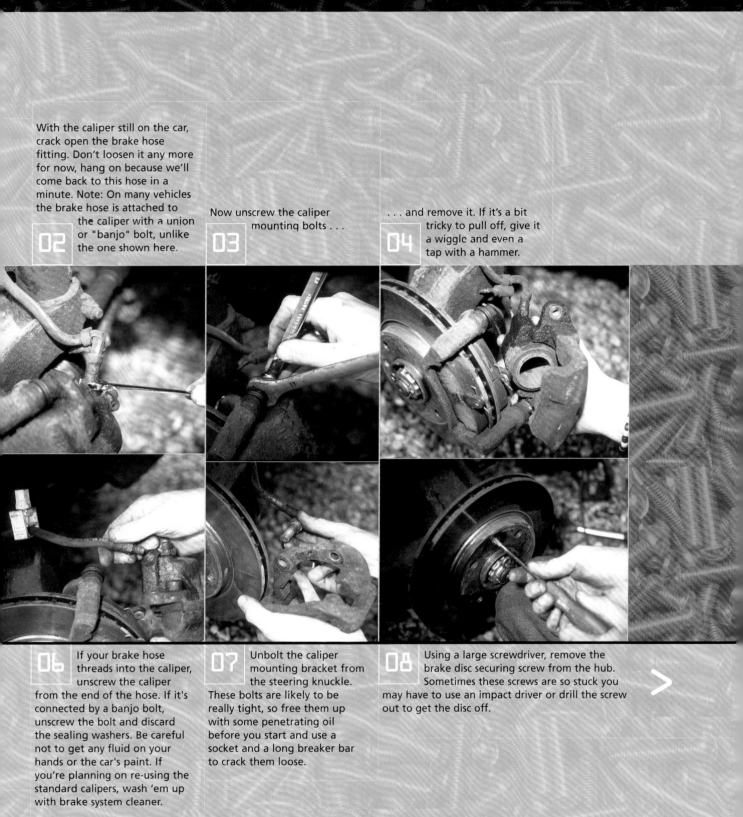

02 With the caliper still on the car, crack open the brake hose fitting. Don't loosen it any more for now, hang on because we'll come back to this hose in a minute. Note: On many vehicles the brake hose is attached to the caliper with a union or "banjo" bolt, unlike the one shown here.

03 Now unscrew the caliper mounting bolts . . .

04 . . . and remove it. If it's a bit tricky to pull off, give it a wiggle and even a tap with a hammer.

06 If your brake hose threads into the caliper, unscrew the caliper from the end of the hose. If it's connected by a banjo bolt, unscrew the bolt and discard the sealing washers. Be careful not to get any fluid on your hands or the car's paint. If you're planning on re-using the standard calipers, wash 'em up with brake system cleaner.

07 Unbolt the caliper mounting bracket from the steering knuckle. These bolts are likely to be really tight, so free them up with some penetrating oil before you start and use a socket and a long breaker bar to crack them loose.

08 Using a large screwdriver, remove the brake disc securing screw from the hub. Sometimes these screws are so stuck you may have to use an impact driver or drill the screw out to get the disc off.

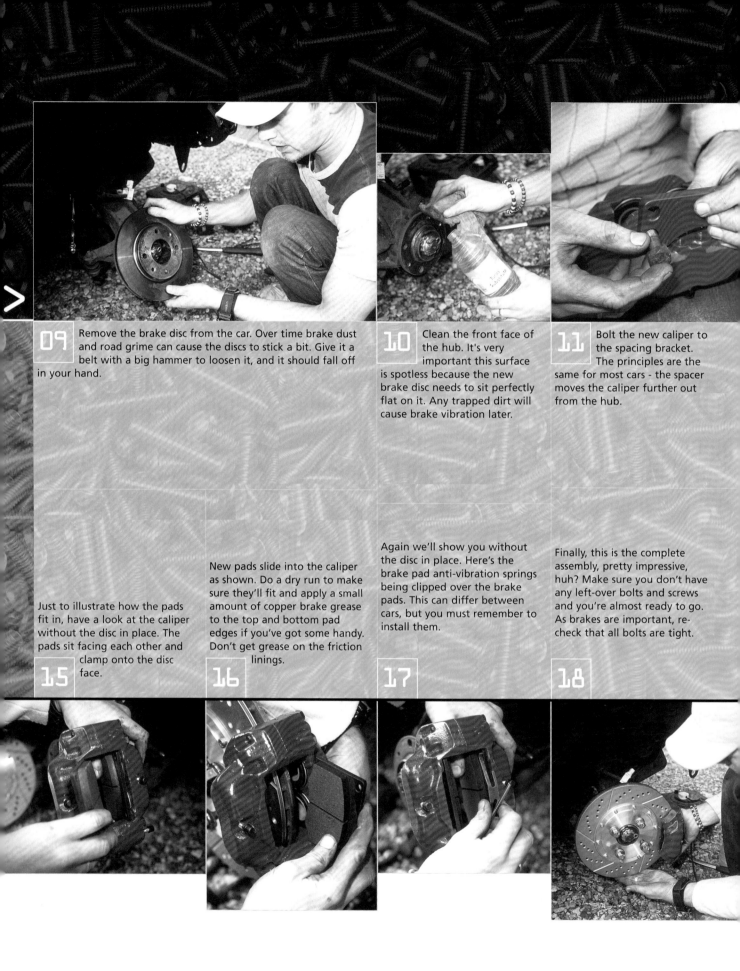

09 Remove the brake disc from the car. Over time brake dust and road grime can cause the discs to stick a bit. Give it a belt with a big hammer to loosen it, and it should fall off in your hand.

10 Clean the front face of the hub. It's very important this surface is spotless because the new brake disc needs to sit perfectly flat on it. Any trapped dirt will cause brake vibration later.

11 Bolt the new caliper to the spacing bracket. The principles are the same for most cars - the spacer moves the caliper further out from the hub.

15 Just to illustrate how the pads fit in, have a look at the caliper without the disc in place. The pads sit facing each other and clamp onto the disc face.

16 New pads slide into the caliper as shown. Do a dry run to make sure they'll fit and apply a small amount of copper brake grease to the top and bottom pad edges if you've got some handy. Don't get grease on the friction linings.

17 Again we'll show you without the disc in place. Here's the brake pad anti-vibration springs being clipped over the brake pads. This can differ between cars, but you must remember to install them.

18 Finally, this is the complete assembly, pretty impressive, huh? Make sure you don't have any left-over bolts and screws and you're almost ready to go. As brakes are important, re-check that all bolts are tight.

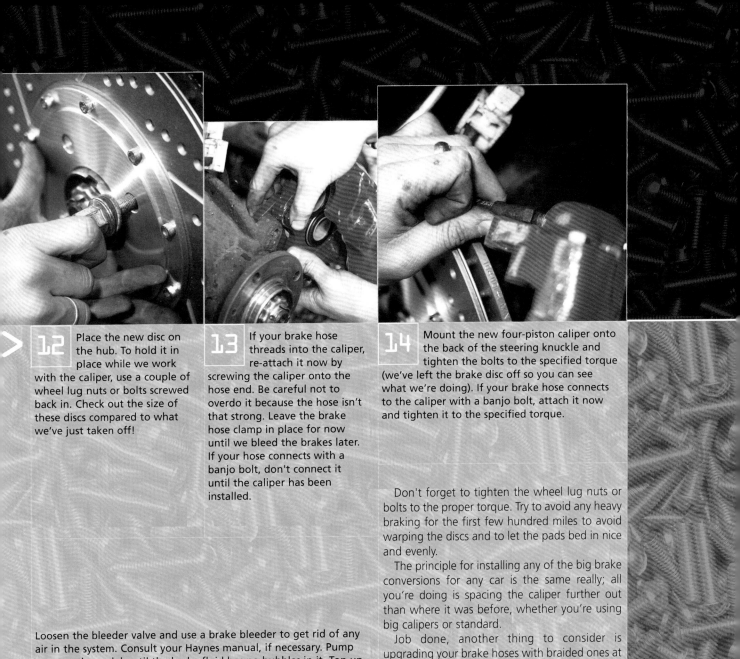

> **12** Place the new disc on the hub. To hold it in place while we work with the caliper, use a couple of wheel lug nuts or bolts screwed back in. Check out the size of these discs compared to what we've just taken off!

13 If your brake hose threads into the caliper, re-attach it now by screwing the caliper onto the hose end. Be careful not to overdo it because the hose isn't that strong. Leave the brake hose clamp in place for now until we bleed the brakes later. If your hose connects with a banjo bolt, don't connect it until the caliper has been installed.

14 Mount the new four-piston caliper onto the back of the steering knuckle and tighten the bolts to the specified torque (we've left the brake disc off so you can see what we're doing). If your brake hose connects to the caliper with a banjo bolt, attach it now and tighten it to the specified torque.

Don't forget to tighten the wheel lug nuts or bolts to the proper torque. Try to avoid any heavy braking for the first few hundred miles to avoid warping the discs and to let the pads bed in nice and evenly.

The principle for installing any of the big brake conversions for any car is the same really; all you're doing is spacing the caliper further out than where it was before, whether you're using big calipers or standard.

Job done, another thing to consider is upgrading your brake hoses with braided ones at the same time - just a thought.

Loosen the bleeder valve and use a brake bleeder to get rid of any air in the system. Consult your Haynes manual, if necessary. Pump the pedal until the brake fluid has no bubbles in it. Top up the fluid reservoir under the hood.

19

Painting calipers

Installing custom rims means that your brakes can be seen by anybody - not a problem usually, but if you've got massive wheels, you really need something good to look at behind them. So if you can't stretch to a big disc conversion or even upgraded pads and discs, why not clean up what you've got there already?

You can spray them or hand paint them, but either way the key is in the prepping. Brakes are dirty things, so make sure they're really clean before you start.

 Tricks n' tips
If you have trouble reassembling your brakes after painting, you probably got carried away and put on too much paint. We found that, once it was fully dry, the excess paint could be trimmed off with a knife.

01 We know you won't necessarily want to hear this, but the best way to paint the calipers is to do some disassembling first. The kits say you don't have to, but trust us - you'll get a much better result from a few minutes extra work. We removed the caliper, pads and mounting bracket, then took off the disc and re-mounted the bracket and caliper on their own. Doing it our way means no risk of paint going on the pads or disc.

02 Get the wire brush out, and attack the rusty old caliper to get rid of all the loose stuff. Wear a mask to avoid inhaling the dust. Squirt on your brake cleaner (some kits come with cleaner), giving the caliper a good dose, and start wiping as soon as possible. Spraying alone will only loosen the muck, and a good scrub is the only answer. If you don't get it spotless, you'll get black streaks in the paint later.

03 Even though we're sure you followed our advice and removed everything except the caliper, there still a bit of masking-up to do, like around the bleeder valves. How much masking you'll need depends on how big a brush you're using, and how steady your hand is.

Painting drums

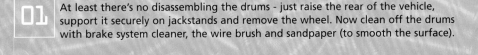

01 At least there's no disassembling the drums - just raise the rear of the vehicle, support it securely on jackstands and remove the wheel. Now clean off the drums with brake system cleaner, the wire brush and sandpaper (to smooth the surface).

02 Mask off the section of drum where the wheel bolts on - you don't want paint there (it could lead to the wheel loosening up later). Trim the mask with a knife to get a neat edge. Also mask off the edge of the brake backing plate (the part that doesn't turn) - you'll hardly see the edge of it when the wheel's on.

03 Painting the drums is much easier than the calipers. One piece of advice - for the drums, use a thicker, better-quality brush than the one you might get in a kit, as you'll get a much smoother paint finish on the drums. Again, two coats of paint seemed like a good idea. Another good idea is to release the parking brake and turn the drum half a turn every so often until the paint is dry.

04 Our paint came in two cans - one paint, one hardener. Pour one into the other, stir, and you then have about four hours max before the paint sets hard in the can. If you're painting calipers and drums, it's best to do all the prep work and be totally ready to start painting at all four corners of the car before you mix the paint.

05 Stick some paper under the brake so it doesn't matter where it ends up. Remember that you only have to paint the parts you'll see when the wheels are on. Once you start painting, it's a job knowing where not to paint. It's best to do more than one coat if you can, we found. Follow the instructions with your kit on how long to wait between coats, but remember the time limit before the paint in the can is useless. Wait until the paints totally dry (like overnight or longer) before reassembling.

10 Engine performance

It's been said: "speed is just a question of how much you want to spend." There's a lot of truth in this statement. People with big incomes and racers with sponsors can afford to build 7-second cars.

But remember also that speed comes with compromise, and an engine built to make 600 horsepower at the racetrack is not going to idle smoothly, get good gas mileage or go 200,000 miles between overhauls. Performance modifications are frequently called "upgrades," but we need to keep in mind that the performance is what's being upgraded, and often you'll have to give up some of the smooth, reliable and economical operation that sport-compact cars are famous for.

So it's best to have a plan for your project, even if you don't have all the money to do everything right away. Don't plan on having a 9-second car that you can still loan to your mom on grocery day – neither one of you will be happy. Most of us will want to build a car that is a compromise: a car that is fast enough to race on the weekend, but is still practical to drive to work every day. Many upgrades, such as an exhaust header, cat-back system, and air-intake tube give "free" horsepower, with the only compromise being more noise (or as we like to call it, "engine music"). These upgrades are also relatively inexpensive and are "no-brainers" for any sport-compact build-up. When you get into nitrous, turbos and superchargers, you'll be spending more money and also getting into more risk of engine damage. Camshafts and cylinder head work will reduce your car's low-speed driveability and frequently decrease your gas mileage. Often, when these modifications are designed to increase high-rpm horsepower, you'll actually lose some low-rpm power.

But enough of the negative. Let's have some fun building your "ultimate" engine!

Modifying exhaust

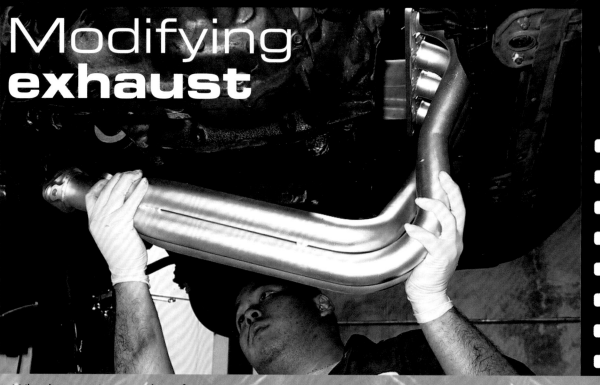

Whether you're a real performance crazy or just after a better sound and look, your car's exhaust system is important. What's lucky for us is that a good performance exhaust system can satisfy both interests.

When you assemble the right package of exhaust components that allow your engine to really breathe, the car's going to sound as good as it performs. It bears repeating here that the exhaust system is one of the few aspects of modifying that gives you the performance you want without any of the drawbacks or compromises that usually come with engine mods. On the contrary, the exhaust work should have no effect on your idling or smooth driveability, and your fuel economy will actually go up, not down! The cool sound is a bonus, too.

Backpressure and flow

Even a stock engine that operates mostly at lower speeds has to get rid of the byproducts of combustion. If it takes fuel and air in, it has to expel those gasses after the reciprocating components have turned the cylinder pressures into work. The exhaust system needs to allow swift exit of those gasses, and any delay or obstruction to that flow can cause engine efficiency to suffer. If there is an exhaust flow problem somewhere in the system, the pressure waves coming out with the gasses can "back up", which can cause the cylinders to work harder to complete their four-stroke cycle. In exhaust terms, this is called backpressure, and getting rid of it is the chief aim of performance exhaust system designers.

We know that the right-sized length of straight pipe, with no catalytic converter and no muffler, would probably offer the least backpressure to the engine, but its not very practical for a street-driven machine. For our purposes, we have two main components between the exhaust valves and the tip of the tailpipe: the header, and everything from the catalytic converter back (which includes the converter, exhaust pipe, silencer, tailpipe and muffler. If we can improve both of these, we 'll have the best street exhaust system possible.

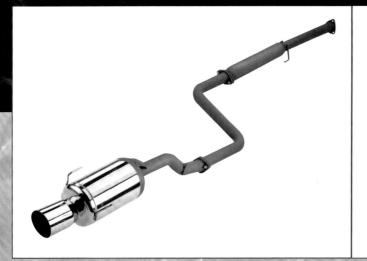
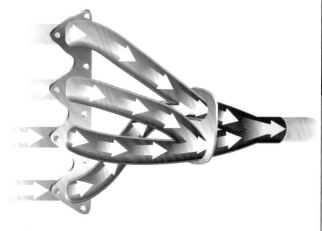

Headers

The first major component of your performance exhaust system is the header, a tubular replacement for your stock cast-iron exhaust manifold. In most OEM sport compact applications the stock exhaust manifold isn't too bad, at least for the needs of your stock economy engine. The exhaust flow needs of an engine go up exponentially with the state of performance "tune." The flow and backpressure needs of a stock engine aren't excessive, especially when the engine spends the bulk of its driving life under 4000 rpm. But what happens when you modify the engine and then take advantage of those modifications by using the high end of the rpm scale? The exhaust system that was once adequate is now restrictive to a great degree.

How much power you make with just the header depends on several factors. On a car with a really restrictive stock exhaust system, particularly an exhaust manifold full of tight bends and twists, the header will make a bigger improvement than on a car with a decent system to start with. What's behind the header can make a big difference as well. If the stock exhaust system includes restrictive converter and muffler designs, small-diameter exhaust pipes and lots of "wrinkle bends" to boot, the header isn't going to have much chance of making a big improvement in performance. A good header on a typical sport compact engine with few other modifications can be expected to make only 3-5 horsepower, depending on how good or bad the stock manifold had been. That's with a stock exhaust system from the header back.

That sounds disappointing, but if we take a case where the engine has numerous modifications yet still has a stock exhaust manifold, the same aftermarket header could gain 10 horsepower. Any of the big "power-adder" modifications, including nitrous oxide and supercharging virtually "require' a header and free-flowing exhaust system to take advantage of their power potential.

Installation of a performance header is quite easy. Most aftermarket headers are made to use all the factory mounting points and braces, in fact, beware of one that doesn't because the header may sound "tinny" when installed if the proper braces aren't connected. Make sure before buying a header that you give the shop your exact year, model and if it's an engine swap or not, especially if it's an imported, used Japanese engine you have swapped in. The location of the oxygen sensor varies among models, and especially between US and JDM installations. Most headers we've seen do not have provisions for bolting on the factory tin heat shield over the header, but we're pretty sure you want to show off your new pipes anyway.

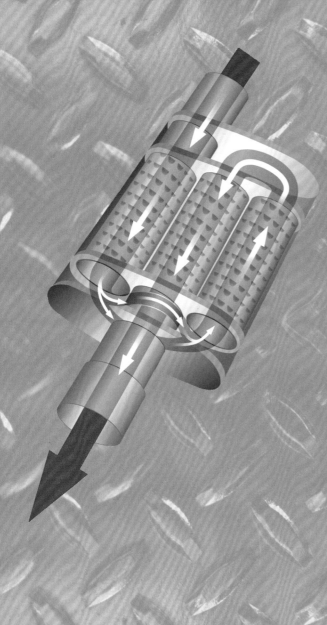

Cat-back system

Once the exhaust gasses leave your efficient new header, they must travel a ways to get out from under your car, and our goal is to make it more of a freeway than a mountain road. First off, there is the catalytic converter, usually bolted right there to the collector of your header. For most of us, this is not an optional component because we have to have it to maintain emissions levels and pass annual or bi-annual state tests. That's fine, because the converter does do a great job of cleaning up engine emissions, and the designs of current converters are more free-flowing than older ones, so the converter isn't something to whine about too much.

Most of the aftermarket exhaust systems are called "cat-back", because they include everything from the converter back. The system may be in several pieces to simplify packaging, shipping and installation, but it should bolt right up to the back flange of your converter and include whatever silencer(s) and muffler(s) are needed. Many of the cat-back systems include the muffler, which on most sport compacts is a part of the car that's very visible, especially when it's a really big one mounted right at the rear bumper.

The muffler you choose has a proportionately large share of both the sound level and backpressure of your total exhaust system. We all like a "good" growl or purr, but a really free-flowing muffler may be loud to the point of annoying. Maybe what we want is like the sound of a Ferrari: purrs at idle; draws attention when accelerating; and at full song its like music. Some muffler companies actually have their mufflers divided into "classes", with one group having the "most aggressive" sound, another a little more subdued, and one that's described s "mellow." The perfect sound is such a subjective thing, you may want to just listen around when you're at events with a lot of sport compacts. When you hear a car whose tone suits you, check for a name on the muffler or ask the owner.

Changing the exhaust of your vehicle for a performance system is one of the modifications with more perks than almost any other. You get increased power, improved fuel economy, the sound that will complement your performance profile, and parts that make your ride look better, too. All that, and there's no real downside or sacrifice as with most engine mods.

Turbochargers

Of the three major "power adders" (nitrous oxide, superchargers, and turbochargers), the turbocharger offers the biggest power potential of all. It's durability and practicality has been proven many times over on production cars and trucks worldwide, whenever extra power was needed without the trouble and expense of fitting the vehicle with a bigger engine.

How it works

There are basically three main elements to a turbocharger: the exhaust turbine; the intake air compressor; and the housing/shaft/bearing assembly that ties the two pressure-related sections together. The job of the turbine is to spin the shaft of the turbocharger. The turbine is composed of an iron housing in which rotates a wheel covered with curved vanes or blades. These blades fit precisely within the turbine housing. When the turbine housing is mounted to an engine's exhaust manifold, the escaping hot exhaust gasses flow through the housing and over the vanes, causing the shaft to spin rapidly. After the exhaust has passed through the turbine it exits through a large pipe, called a downpipe, to the rest of the vehicle's exhaust system.

Within the compressor housing is another wheel with vanes. Since both wheels are connected to a common shaft, the intake wheel spins at the same speed as the turbine, so the compressor draws intake air in where the rapidly spinning wheel blows the air into the engine's intake side. The more load there is on the engine, the more the turbocharger works to give the engine horsepower. As the engine goes faster, it makes more exhaust, which drives the turbocharger faster, which makes the engine produce more power.

The unit in the center of this turbo "sandwich" has the important function of reliably passing the power between the two housings.

Advantages and disadvantages of the turbocharger

Compared to the supercharger, the turbo is smaller, lighter, quieter, and puts less direct load on the engine. The "not at all times" boost of the turbocharger is an advantage for fuel economy, operating noise level and driveability, but in some cases may be a drawback when compared to the mechanical supercharger. While a supercharger adds boost in relation to rpm (i.e. the faster the engine goes, the faster the blower pumps) it runs up against physical limitations eventually and can't pump any more. In order to make any more boost with a supercharger, you have to change the pulleys or gearing that drives it.

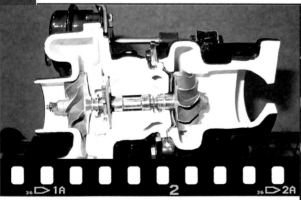

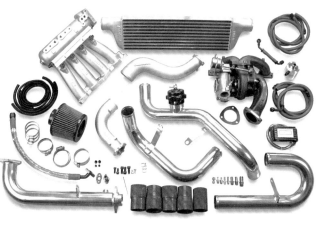

The limitation for the blower-equipped car is in top end performance, and this is where the turbocharger has the distinct advantage. The turbocharger is much more customizable, with various wheels and housings available to suit whatever the intended engine or purpose. Boost can be made to come in early, or come in later at a higher boost level. The same basic turbocharger can be used for street use or modified to make more boost than your engine can live with! Such customizing, called sizing, should be done by an experienced turbo shop that can select the exact right components for your engine size and power requirements.

Intercoolers

One of the more serious disadvantages of the turbocharger is the heat it will put into the intake air charge, which is why most successful turbo systems utilize an intercooler to combat this. Air coming from turbochargers tends to be hot because it's compressed and also because of its close proximity to the exhaust. The most effective and most common method of dealing with air temperature in the intact tract on a turbocharged car is an intercooler. This is a honeycomb affair much like a radiator, usually mounted out in front in an opening below the bumper, where cooler air is found. The boosted air from the compressor is ducted through pipes and into the intercooler, and then to the intake of the engine. Thus, the cooler is between the compressor and the engine, so it's called an inter-cooler. Since cooler air produces more power, intercoolers are very popular.

Turbocharging kits

A complete turbocharging kit from most reputable aftermarket companies will include virtually everything you need for your application, with a cost from $2500 to $4500 for everything. If the price of a good turbo kit sounds expensive, just compare turbocharging to other methods of upping your horsepower. A turbo could be your best horsepower-per-dollar investment. Installation will vary widely between different model cars and turbo kits. Can you install a turbo kit at home with just a standard tool set? A definite maybe. You'll really have to do some research to answer that.

Engine performance

Superchargers

An internal-combustion engine operates like a compressor. It sucks in a combustible mixture of air and fuel, ignites it and uses the resulting pressure to push the piston up and down and make the crankshaft deliver the power to the transaxle. For purposes of this Section, the key word in that very simplistic explanation is "sucks."

A turbocharger, is similar in function to the supercharger, but differs in that the turbo is driven by exhaust gasses, rather than by mechanical means. A supercharger is driven by the engine, either with gears, chains or belts, so there is direct correlation between the engine speed and the boost produced by the supercharger. While the turbocharger may have the upper hand when you're talking about all-out high-rpm performance on the track, the supercharger shines at improving street performance almost from idle speed on up. Low-rpm power is very helpful for "across-the-intersection" performance, where a turbocharger is probably just "spinning up" and not yet able to provide much power.

Boost

Boost, is any pressure beyond 14.7 psi (normal atmospheric pressure). So, if you have a turbo or supercharger that is making 14.7 psi of boost, then you have effectively doubled normal atmospheric pressure or added another "atmosphere" of pressure. Twice that amount of boost and you've added two atmospheres, and you command the bridge of a rocket ship!

But, we must be realistic about boost levels. There is a finite limit to the boost your blower can make, and probably a much lower limit of how much boost your engine can take, regardless of how the boost was generated. You'll find most street supercharger kits are limited to 5-7 psi, to make some power while working well on a basically stock engine that sips pump gasoline. Some kits on the market have optional pulleys that will spin the blower faster for more boost, but, as with any power adder, you can only go so far in increasing cylinder pressure before you have to make serious modifications to strengthen the engine (forged pistons, aftermarket connecting rods, etc.).

Heat and detonation

Knock, ping, pre-ignition and detonation are all terms that describe abnormal combustion in an engine and usually describe big trouble. An increase in boost also increases the combustion chamber temperature and pressure, often leading to these problems.

The two main factors in detonation and its control are heat and timing. Ignition timing does offer some ways to deal with detonation. Supercharged engines generally "like" more timing, especially initial advance, but once you are making full boost and the vehicle is under load, too much ignition advance can bring on the "death rattle" of detonation that you don't want to hear. How much advance your application can handle is a trial-error-experience thing, but if you are using a production blower kit from a known manufacturer, these tests have already been made and some kind of timing control program should be included.

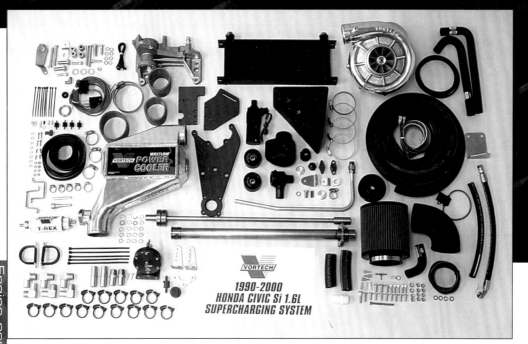

Blower kits

There are several types of small superchargers used in kits designed for compact-car engines. The two main types of supercharger design you will see are the roots type or "positive displacement", and the centrifugal design.

The roots-type mechanical blower uses a pair of rotors that turn inside a housing. As the rotors turn, they capture a certain amount of air and propel it to the inner circumference of the case and out to the intake manifold. Each time they turn around, they capture air, hence the "positive" description. The benefit of this type is that it starts making boost at very low rpm.

A variation of this type of blower is the screw-type. These have two rotors with helically-wound vanes that, as the name implies, look like two giant screws. When the two screws mesh together (there is a male and a female rotor), the air is actually compressed between the screws.

The other major type of bolt-on supercharger is the centrifugal design. From a quick examination, the centrifugal blower looks just like the compressor half of a turbocharger, being a multi-vaned wheel within a scroll-type housing. Unlike a turbocharger, this type of blower isn't driven by exhaust but by a mechanical drive from the engine, usually a belt. These types of blowers are not "positive-displacement", and thus do not necessarily make their boost down on the low end, but have plenty of air-movement potential when they are spinning rapidly.

Most reputable blower kits have everything you need to install the system and use it reliably. Contents include the blower, intake manifold (if needed), belt, mounting hardware, instructions and some type of electronic gear to control fuel delivery and/or ignition timing. Some kits use a larger-than-stock fuel pump that replaces your in-tank pump, and a special fuel-pressure regulator that may be boost-sensitive. Other included components could be hoses, wire harnesses, air intake, an intercooler, and parts that relocate items in the engine compartment to provide room for the blower.

Nitrous Oxide

Nitrous oxide as a horsepower source can appear to be a miracle or a curse, depending on your experience. It's a fact that nitrous oxide (N2O) is the simplest, quickest and cheapest way to gain a large horsepower boost in your car. It's often referred to as the "liquid supercharger." But, just like in children's fairy tales, you court disaster if you don't follow the rules that come with the magic potion.

Nitrous oxide is an odorless, colorless gas that doesn't actually burn. It carries oxygen that allows your engine to burn extra fuel. When you inject nitrous oxide and gasoline at the same time and in the proper proportions at full throttle, you'll get a kick in the butt as if you were instantly driving a car with an engine twice as big!

Done right, a nitrous kit is one of the best horsepower-per-dollar investments you can make, and is the most popular power-adder for small engines. Kits are available that add from 50 hp to 300hp, because the more nitrous and fuel you add, the more power the engine makes . . . right up to the point where the engine comes apart. In the end, it is the durability of the engine itself that determines how much nitrous you can run. As tempting as it is to just keep putting bigger nitrous jets in your engine for more power, you need to do your research first to find the limits of your engine and what can be done to protect against damage done by a little too much nitrous fun.

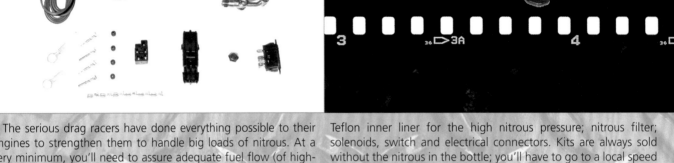

The serious drag racers have done everything possible to their engines to strengthen them to handle big loads of nitrous. At a very minimum, you'll need to assure adequate fuel flow (of high-octane fuel) and probably retard the ignition timing. Many racers use electronic retard boxes that allow you to retard the ignition timing from the driver's seat. Usually, you'll need to put in a more capable ignition system than the stock one (with colder spark plugs), and it's wise to add a low-fuel-pressure shut-off switch – this device will save your engine if your fuel pressure ever drops too low while you're "on the bottle." If you're going to push much past a 60-horsepower system, and certainly if you're into the 100 horsepower-and-up category, you'll need to consider internal engine upgrades to handle the additional horsepower. Generally, forged pistons are used with these higher-horsepower kits, and most stock engines generally do not have forged pistons. So, if you're planning to have your engine live very long at these high-horsepower levels, you'll need to spend some money upgrading the "bottom end" of the engine.

Nitrous kits

The basic street-use nitrous kit consists of: a nitrous bottle, usually one that holds 10 pounds of liquid nitrous oxide; bottle mounting brackets and hardware; fuel and nitrous jets; high pressure lines, usually braided-stainless covered AN–type with a Teflon inner liner for the high nitrous pressure; nitrous filter; solenoids, switch and electrical connectors. Kits are always sold without the nitrous in the bottle; you'll have to go to a local speed shop to have the bottle filled.

There are two basic types of nitrous kits that differ in how the nitrous and fuel are delivered. The "dry" system has a nozzle and solenoid only for the nitrous, and this nozzle may be almost anywhere in the intake system, behind the MAF (Mass Air Flow) sensor (if equipped) and ahead of the throttle body is typical. The dry systems are designed for factory fuel-injected engines. To deliver the extra fuel to go with the nitrous, the stock system fuel pressure is raised during the WOT (wide-open throttle) period of nitrous usage, and reverts back to the stock fuel pressure in all other driving conditions.

The "wet" nitrous systems have solenoids for both nitrous and fuel, both of which are turned on when the nitrous system is activated. The fuel and nitrous in a simple system are plumbed into a single injector that merges the two, the gasoline and its oxidizer, as they enter the engine. More sophisticated wet systems may have an nitrous nozzle for each cylinder of the engine, and one or more fuel nozzles mounted separately. In high-output racing applications, individually feeding each cylinder allows for tuning each cylinder separately for fine control. The extra nozzles also mean there is ample supply of nitrous and fuel for very high-horsepower installations.

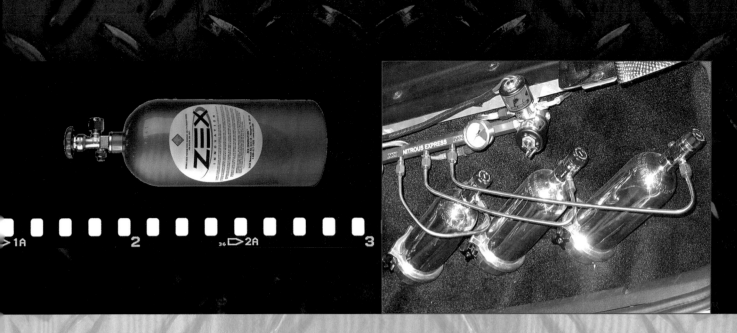

There are all sorts of "bells and whistles" available as extra features on most of the nitrous kits on the market today. There are varying designs of nozzles, different electronic controls that work with your factory computer to control timing and fuel delivery, optional bottle covers, bottle warmers, remote shut-off valves for the bottle, PCM piggybacks that pull back the ignition timing under nitrous use, and "staged" nitrous kits that delivery a certain amount during launch, then a little more, and the full blast for the top end.

How does it work?

In basic terms for a simple nitrous system, the tank or bottle of nitrous oxide is mounted in the trunk and plumbed up to the engine compartment with a high-pressure line. This is connected to an electric solenoid, from which nitrous (still a liquid at this point) can flow to the nozzle attached to your intake system. If you turn on an "arming" switch in the car, battery voltage is available to another switch that is a button-type located on the steering wheel, dash or shifter. Push that button when you're hard on the throttle and the solenoid releases nitrous oxide, which passes through a sized jet or orifice and turns into a vapor, to mix with your vaporized fuel in the engine. The extra fuel admitted at the same time as the N2O is easily oxidized and creates enough cylinder pressure to add 50, 75, 100 or more horsepower in an instant. So, do some research, talk to some tuners and then decide if going on the bottle is the right choice for you.

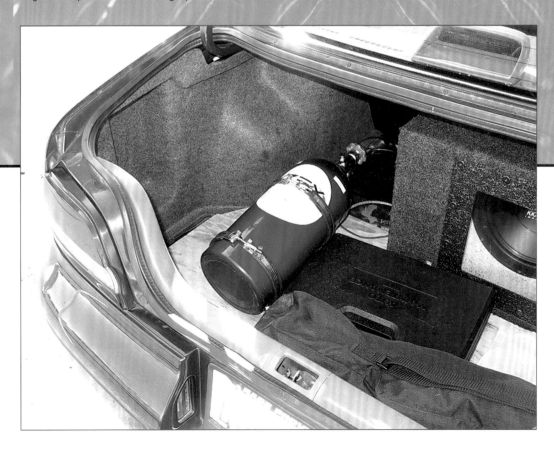

Induction systems

Performance air intakes are available in two basic forms, the "short ram" and the "cold-air" intake. In each case the idea is to get more - and colder - air into your engine. Oh, and they look nice too!

The least expensive and easiest intakes to install are the short ram types. You'll spend longer getting the stock air filter box and inlet tube out than installing the short ram. If you have any doubts about removing the stock components, consult the Haynes repair manual for your car. Short-ram intakes place the new filter relatively close to the engine, and modifications to the engine or body are rarely necessary. Most short-ram installations take only a half-hour to install and may be good for 4 to 8 horsepower, depending on the application. And of course, they look and sound really nice.

For purposes of performance, the colder the air the more power you make. Cold-air intake systems are one of the most-widely-installed bolt-on performance improvements in the sport compact world. They are usually the first modification made to an import engine, and have become so common that people want one on their car whether it makes any more power or not. Since they are available in a variety of flashy finishes, such as anodizing in blue, red, or purple, or a polish finish on the metal tube, they make the most power statement that your engine is "modified." When the tube made of aluminum or stainless steel and is polished, "wow" factor is upped considerably, and these finishes easy to take care of.

Aside from the "look" of the intake, the cold-air packa is important because it is longer, reaching down to pick colder air from below the grille or in the car's fenderwell, rath than the hotter engine compartment air.

Aftermarket cold air intake systems can be worth 8 to horsepower, depending on the design and the quality of the filter included with it. Obviously, a cold air intake is going need some bends in order to reach the cold air, but if there too many bends or bends made too sharp, the horsepov gain from the colder air could be offset by a reduction airflow.

01 The simplest modification you can make to your stock air induction system is to open the factory airbox, lift out the OEM paper air filter, and replace it with a high-flow aftermarket filter with a pleated-cotton lifetime filter - these can be reused over and over by washing them, then treating them with a special oil

02 A typical aftermarket bolt-on induction mod is to install a short ram, which is a new, large diameter pipe with an aftermarket filter - these flow much better than stock, though they do make the engine noisier - this installation doesn't allow the best design, since there is a sharp bend near the throttle body that could slow down the airflow

03 If you do run a short ram, try to find one with a sheetmetal "dam" included in the kit - this can keep hot air radiating from the engine isolated from the filter

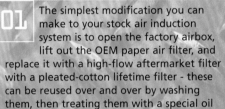

This Typhoon ram from K&N is typical of cold air induction systems, shown here with a "rain hood" that slips over the filter for inclement weather - the junction of the two pipes is a spot where the lower pipe can be removed to attach the filter there inside the engine compartment, making a short ram out of a long ram, another way of preparing for winter driving **04**

05 Longer rams can pick up colder undercar air - this installation on a Ford Focus looks like a short ram, but the pipe ducks down under the battery box and gets air from under the bumper

Computers and chips

There's no sense in putting on lots of goodies that say Go, Go, Go, when your computer is saying No, No, No!

In the world of cars, the computer has become a true boon. As the engineers really got into it over the years, they kept finding new uses for computer control - everything from engine system to air conditioning. Where once there may have been crude information sampling, now cars have higher-performance computers, more sensors, more sensitive sensors, and impressive programming. Some cars have a main computer with one or more "sub-computers" that process data and transfer information between other on-board systems and the main computer. Automatic climate-control systems, where you set the temperature you want in the car and the systems provide just the right level of heat or cooling to maintain that temperature, are a perfect application for a sub-computer.

Of course, the computer and its associated components make a sophisticated and complex automotive system. The information provided here is only a basic description, and you should consult with experts before making changes to your computer system. You can find specific information for your car by buying a Haynes repair manual.

Engine management basics

Automotive computer systems consist of an onboard computer, referred to by the factory as the Powertrain Control Module (PCM) or Engine Control Module (ECM), and information sensors, which monitor various functions of the engine and send data to the PCM. Based on the data and the information programmed into the computer's memory, the PCM generates output signals to control various engine functions via control relays, solenoids and other output actuators.

The PCM is the "brain" of the electronically controlled fuel and emissions system, and is specifically calibrated to optimize the performance, emissions, fuel economy and driveability of one specific vehicle/engine/transaxle/ accessories package in one make/model/year of vehicle.

Computer codes keep track

You may heard of the term OBD or OBD-II. This means On-Board Diagnostics and refers to the ability to retrieve information from the PCM about the performance characteristics and running condition of all the sensors and actuators in the engine management system. This is invaluable information in diagnosing engine problems. The PCM will illuminate the CHECK ENGINE light (also called the Malfunction Indicator Light) on the dash if it recognizes a component fault.

So, if your dashboard warning light comes on you know the computer has spotted something it doesn't like. To then figure out what it has found, you (or your mechanic) need to access the diagnostic code that the computer has stored in its memory for that fault. On some vehicles, getting these codes is an easy in-the-driveway job. On others, it takes an expensive "scan tool". Your Haynes manual will give specific information for your make and model.

Add-on chips and computers

For improved performance, many enthusiasts upgrade their computer chips, or replace their entire computer. The advantages are increased fuel flow, an improved ignition advance curve and higher revving capability. While replacing computer components can provide substantial performance gains when combined with other engine upgrades, they also have their downside. Perhaps most importantly, replacing any original-equipment computer components can void your warranty or cause you to fail an emissions inspection. Nevertheless, if you are seriously into modifying your engine, at some point you will have to consider dealing with the computer.

Information sensors

Although most cars have many more sensors than those listed here, these are the sensors that provide the most important information about the engine running condition:

Oxygen sensor (O2S) - The O2S generates a voltage signal that varies with the difference between the oxygen content of the exhaust and the oxygen in the surrounding air. The PCM reads this signal voltage as a too-lean, too-rich or just-right fuel mixture. The PCM then varies the fuel output at the injectors to bring the engine into perfect fuel trim. This fuel-mixture monitoring happens about 100 times per second.

Engine Coolant Temperature (ECT) sensor - The ECT sensor monitors engine coolant temperature and sends the PCM a voltage signal that affects PCM control of the fuel mixture and ignition timing. This is important information for the PCM, since the fuel mixture must be richer when the engine is cold.

Intake Air Temperature (IAT) sensor - The IAT provides the PCM with intake air temperature information. The PCM uses this information to control fuel flow, ignition timing, and EGR system operation. Generally, when the intake air is colder, the fuel mixture must be richer.

Throttle Position Sensor (TPS) - The TPS senses throttle movement and position, then transmits a voltage signal to the PCM. This signal enables the PCM to determine when the throttle is closed, in a cruise position, or wide open. For example, in cruising conditions, the fuel mixture can be leaner than when it is at wide-open throttle.

Mass Airflow Meter (MAF) – The MAF sensor measures the mass of the intake air by detecting volume and weight of the air from samples passing over a hot wire element. The PCM uses this information to choose the precise amount of fuel to inject.

Output actuators

The PCM uses the information from the information sensors to figure out what changes it needs to make to the engine running condition so the engine will run perfectly for the existing conditions. The PCM does this by using output actuators that it can adjust. The following are the basic actuators used by the PCM – most cars have more:

Fuel injectors - The PCM opens the fuel injectors individually. On most modern engines, the system is called SFI (Sequential Fuel Injection), in which the injectors are opened sequentially according to the firing order of the cylinders. The PCM also controls the time the injector is open, called the "pulse width." The pulse width of the injector (measured in milliseconds) determines the amount of fuel delivered, which can be varied to make for richer or leaner running.

Ignition module or igniter - The ignition module or igniter triggers the ignition coil and determines proper spark advance based on inputs to the PCM.

Idle air control (IAC) valve - The IAC valve controls the amount of air to bypass the throttle plate when the throttle valve is closed or at idle position. By varying this air bypass, the PCM can raise or lower the idle speed.

Valvetrain modifications

When you run fast, you breathe harder – so does your engine. To make more power, an engine must "inhale" more air and "exhale" more exhaust. To make this happen, you can open the intake and exhaust valves more, leave them open longer and/or enlarge the "ports" (passages in the cylinder head where the air and exhaust flow). These cylinder head and valvetrain modifications can seem complicated, but understanding them is essential. Mistakes here can cost you power or even an engine overhaul.

The modifications discussed here usually come after all the other bolt-ons have not gained you the power you're after. Most of these modifications require going inside the engine, which is not a place an amateur should go alone. You'll need to find a reputable tuner who knows your particular engine inside and out.

Cam sprockets

On Double Overhead Cam (DOHC) engines, cam gears (more properly, sprockets), are a very popular modification. They not only look great, they can actually provide a few extra horsepower when set up properly. While dialing in the degrees on the sprockets is relatively easy, you will need to know the precise number of degrees to advance or retard each cam in your particular application. The correct "degreeing in" specifications are determined by the engine type, level of modification and type of power you're after (low-rpm or high-rpm). Because of this complexity, it's best to ask a tuner who's familiar with your type of engine.

Installing cam gears means you'll have to remove, then reinstall the timing belt. This procedure is best left to a qualified technician, since any slight error could cause you to bend your valves from valve-to-piston contact. At the very least, get the Haynes manual for your particular vehicle and follow the procedure carefully. Be sure to rotate the engine through two turns by hand after the belt is back on – this way you'll identify any problems before turning the key to the sound of grinding metal if you make a mistake..

Camshafts

Installing camshafts is an expensive and precise piece of work. The most important work comes before any tools come out. You, consulting with your tuner, need to pick out the best cam for your car and driving style. Basically, you also need to figure out how much "cam" you really want or need.

Stock camshafts are designed as a compromise to consider economy, emissions, low-end torque and good idling and driveability. The performance camshaft lifts the valves higher (lift), keeps them open longer (duration) and is designed mainly to produce more horsepower. A performance camshaft usually makes its gains at mid-to-higher rpm and sacrifices some low-rpm torque. The hotter the cam, the more pronounced these attributes become. A cam design that is advertised for power between 3000 and 8000 rpm won't start feeling really good until that rpm" band" is reached. Aftermarket cams for sport compact cars are usually offered in "Stages" of performance. A typical Stage 1 cam might have a little higher lift than stock and a little longer duration. It would still keep an excellent idle and work from idle or 1000 rpm up. A Stage II cam would be

hotter in all specs (with a band from 3000 to 7000 rpm) and have a slightly rough idle (maybe 750 rpm). A Stage III cam would feature serious lift, duration and overlap and make its power from 5000 to 8000 rpm. The hotter the cam specs, the worse the idle, low-end performance and fuel economy is going to be, but the more top-end horsepower you'll make.

Valve springs

When you install a "bigger" camshaft, you'll usually want to install better valve springs. High-performance valve springs allow the valves to open further without the springs binding and also are stronger to prevent valve "float." Valve float occurs when the engine is at very high rpm and the inertia of the valve is too much for the spring to handle. So the valves actually lose contact with the lifter or cam follower and can make contact with the piston, bending the valve and/or damaging the piston. For this reason, high-performance valve springs are recommended whenever you change the camshaft or make other modifications to extend the rpm range beyond stock.

Cylinder head work

The most basic of cylinder head work is a valve job, which will assure the valve seats and guides are in good condition. This is very important for an engine running at high rpm, and is essential if your engine has very many miles on it. If you're planning to install camshafts, make sure you get a good valve job, preferably a three-angle valve job from a performance machine shop. You won't see much performance gain from this work: it's insurance against damage on an engine that will be pushed to the limits.

Cylinder head porting is a specialized art and science that is practiced by performance machine shops. Don't try it at home! Novices can damage the cylinder head or actually reduce airflow. A performance machine shop has all the right equipment for the job, including a flow-bench. With the aid of a flow-bench, the performance machine shop can "open up" your heads for maximum performance. Be sure to find a shop that specializes in your type of engine.

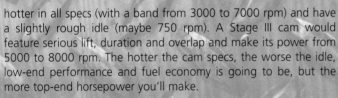
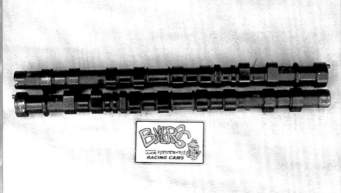

Engine performance

Ignition system

Higher engine speeds put an increased load on stock ignition systems, but modifications that lead to increased cylinder pressure can really put out the fire. If you install higher-compression pistons - always a good move for increased engine performance - the spark plugs need a lot more zap to light off a mixture that is packed tighter than ever. The denser the mixture, the harder it is for a spark to jump the plug's electrodes, like swimming through wet concrete. Other major power-adders such as nitrous oxide, supercharging or turbocharging also create much higher cylinder pressures and require ignition improvements to handle this.

Aftermarket ignition coils

The typical aftermarket coil is capable of making more secondary voltage than the stock coil, so the spark can jump the gap even at extreme cylinder pressures. An aftermarket coil won't add any horsepower and won't improve your fuel economy, but it could eliminate some misfire problems in the upper rpm range, and that will become more important as you add other modifications to your engine.

High-performance plug wires

Most aftermarket performance wires use a very fine spiral wire wound around a magnetic core and wrapped in silicone jacketing. They are available in thicker-than-stock diameters to handle more current flow. Some import cars have stock plug wires as skinny as 5 or 6mm, while aftermarket wires are offered in 8mm, 8.5mm and even 9mm for racing applications. Good aftermarket wires also come with thicker boots, which is important, since the boot-to-plug contact area is a frequent source of voltage leaking to ground. Wires in the 8mm range are big enough to handle the spark of most street-modified cars, and the bigger wires are really only necessary for racing.

Spark plugs

The final links in the ignition system's chain-of-command are the spark plugs, the front-line combat troops. As with other ignition modifications, don't expect to make big gains in power or mileage by switching spark plugs. New spark plugs could bring back 5 or 10 lost horsepower if you haven't replaced them in a while and they're really worn. But, most aftermarket plugs can't really make new horsepower the engine didn't have before.

11 Safety First

Regardless of how enthusiastic you may be about getting on with the job at hand, take the time to ensure that your safety is not jeopardized. A moment's lack of attention can result in an accident, as can failure to observe certain simple safety precautions. The possibility of an accident will always exist, and the following points should not be considered a comprehensive list of all dangers. Rather, they are intended to make you aware of the risks and to encourage a safety conscious approach to all work you carry out on your vehicle.

Essential DOs and DON'Ts

DON'T rely on a jack when working under the vehicle. Always use approved jackstands to support the weight of the vehicle and place them under the recommended lift or support points.
DON'T attempt to loosen extremely tight fasteners (i.e. wheel lug nuts) while the vehicle is on a jack - it may fall.
DON'T start the engine without first making sure that the transmission is in Neutral (or Park where applicable) and the parking brake is set.
DON'T remove the radiator cap from a hot cooling system - let it cool or cover it with a cloth and release the pressure gradually.
DON'T attempt to drain the engine oil until you are sure it has cooled to the point that it will not burn you.
DON'T touch any part of the engine or exhaust system until it has cooled sufficiently to avoid burns.
DON'T siphon toxic liquids such as gasoline, antifreeze and brake fluid by mouth, or allow them to remain on your skin.
DON'T inhale brake lining dust - it is potentially hazardous (see **Asbestos**).
DON'T allow spilled oil or grease to remain on the floor - wipe it up before someone slips on it.
DON'T use loose fitting wrenches or other tools which may slip and cause injury.
DON'T push on wrenches when loosening or tightening nuts or bolts. Always try to pull the wrench toward you. If the situation calls for pushing the wrench away, push with an open hand to avoid scraped knuckles if the wrench should slip.
DON'T attempt to lift a heavy component alone - get someone to help you.
DON'T rush or take unsafe shortcuts to finish a job.
DON'T allow children or animals in or around the vehicle while you are working on it.
DO wear eye protection when using power tools such as a drill, sander, bench grinder, etc. and when working under a vehicle.
DO keep loose clothing and long hair well out of the way of moving parts.
DO make sure that any hoist used has a safe working load rating adequate for the job.
DO get someone to check on you periodically when working alone on a vehicle.
DO carry out work in a logical sequence and make sure that everything is correctly assembled and tightened.
DO keep chemicals and fluids tightly capped and out of the reach of children and pets.
DO remember that your vehicle's safety affects that of yourself and others. If in doubt on any point, get professional advice.

Steering, suspension and brakes
These systems are essential to driving safety, so make sure you have a qualified shop or individual check your work. Also, compressed suspension springs can cause injury if released suddenly - be sure to use a spring compressor.

Airbag
Airbags are explosive devices that can cause injury if they deploy while you're working on the car. Follow the manufacturer's instructions to disable the airbag whenever you're working in the vicinity of airbag components.

Asbestos
Certain friction, insulating, sealing, and other products - such as brake linings, brake bands, clutch linings, torque converters, gaskets, etc. - may contain asbestos or other hazardous friction material. Extreme care must be taken to avoid inhalation of dust from such products, since it is hazardous to health. If in doubt, assume that they are harmful.

Fire
Remember at all times that gasoline is highly flammable. Never smoke or have any kind of open flame around when working on a vehicle. But the risk does not end there. A spark caused by an electrical short circuit, by two metal surfaces contacting each other, by a tool falling on concrete, or even by static electricity built up in your body under certain conditions, can ignite gasoline vapors, which in a confined space are highly explosive. Do not, under any circumstances, use gasoline for cleaning parts. Use an approved safety solvent.

Always disconnect the battery ground (-) cable at the battery before working on any part of the fuel system or electrical system. Never risk spilling fuel on a hot engine or exhaust component. It is strongly recommended that a fire extinguisher suitable for use on fuel and electrical fires be kept handy in the garage or workshop at all times. Never try to extinguish a fuel or electrical fire with water.

Fumes
Certain fumes are highly toxic and can quickly cause unconsciousness and even death if inhaled to any extent. Gasoline vapor falls into this category, as do the vapors from some cleaning solvents. Any draining or pouring of such volatile fluids should be done in a well ventilated area.

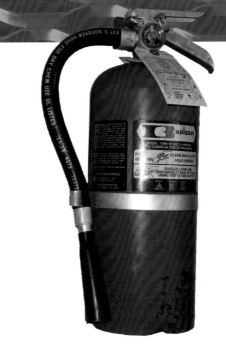

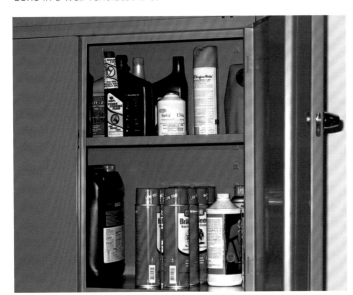

When using cleaning fluids and solvents, read the instructions on the container carefully. Never use materials from unmarked containers.

Never run the engine in an enclosed space, such as a garage. Exhaust fumes contain carbon monoxide, which is extremely poisonous. If you need to run the engine, always do so in the open air, or at least have the rear of the vehicle outside the work area.

The battery
Never create a spark or allow a bare light bulb near a battery. They normally give off a certain amount of hydrogen gas, which is highly explosive.

Always disconnect the battery ground (-) cable at the battery before working on the fuel or electrical systems.

If possible, loosen the filler caps or cover when charging the battery from an external source (this does not apply to sealed or maintenance-free batteries). Do not charge at an excessive rate or the battery may burst.

Take care when adding water to a non maintenance-free battery and when carrying a battery. The electrolyte, even when diluted, is very corrosive and should not be allowed to contact clothing or skin.

Always wear eye protection when cleaning the battery to prevent the caustic deposits from entering your eyes.

Household current
When using an electric power tool, inspection light, etc., which operates on household current, always make sure that the tool is correctly connected to its plug and that, where necessary, it is properly grounded. Do not use such items in damp conditions and, again, do not create a spark or apply excessive heat in the vicinity of fuel or fuel vapor.

Secondary ignition system voltage
A severe electric shock can result from touching certain parts of the ignition system (such as the spark plug wires) when the engine is running or being cranked, particularly if components are damp or the insulation is defective. In the case of an electronic ignition system, the secondary system voltage is much higher and could prove fatal.

Source List

APC
Gearshift knob, handbrake lever/boot, gauge faces
22324 Temescal Canyon Rd.
Corona, CA 92883
(909) 898-9840
www.4apc.net

Focal America, Inc.
ORCA Design and Manufacturing Corp.
Speakers
1531 Lookout Dr.
Agoura, CA 91301
(818) 707-1629
www.focal-america.com

Grant Products
Steering wheel, steering wheel styling ring
700 Allen Ave.
Glendale, CA 91201
(818) 247-2910
www.grantproducts.com

GReddy Performance Products
Turbocharger
9 Vanderbilt
Irvine, CA 92618
(949) 588-8300
www.greddy.com

Grille Craft
Grille
11651 Prairie Ave.
Hawthorne, CA 90250
(310) 970-0300
www.grillecraft.com

JL Audio, Inc.
Amplifiers/subwoofers
10369 N. Commerce Parkway
Miramar, FL 33025
(954) 443-1100
www.jlaudio.com

K&N Engineering
Induction systems
PO BOX 1329
Riverside, CA 92502
(888) 949-1832
www.knfilters.com

Neuspeed/Automotive
Performance Systems/Neumann Distribution
Strut brace, stabilizer bars
3300 Corte Malpaso
Camarillo, CA 93012
(800) 423-3623 Toll free
(805) 388-7171 Technical
(805) 388-8111 Distribution
www.neuspeed.com

SAVV Mobile Multimedia
Mobile video
15348 Garfield Ave.
Paramount, CA 90723
(562) 529-7700
www.savv.com

Vortech Engineering
Superchargers
1650 Pacific Ave
Channel Islands, CA 93033
(805) 247-0226
www.vortechsuperchargers.com

Wings West
Rear wing/body kit
845 West 16th St.
Newport Beach, CA 92663
(949) 722-6114
www.wingswest.com

Woodview
Interior trim kit
5670 Timberlea Blvd.
Mississauga, Ontario, Canada
L4W 4M6
(800) 797-DASH (3274)
www.woodcorp.com

A special thanks to:

- APC (American Products Company) for supplying many of the custom car photos seen throughout this book.
- Hot Import Nights (Vision Events) and Kevin McDonald for access to their great shows.
- Lucette Nicoll of Nicoll Public Relations for arranging much of the in-vehicle entertainment content.
- Ernie, Scott, Carlos and Jose at Wings West for the body kit help and photo shoot.
- Greg Woo and Aaron Neumann at Neuspeed for their technical assistance and help with certain photos.
- Jaime and Hector at Street Sound Plus, Thousand Oaks for all the Mobile Entertainment installations.

HAYNES REPAIR MANUALS

ACURA
- *12020 **Integra** '86 thru '89 & **Legend** '86 thru '90
- 12021 **Integra** '90 thru '93 & **Legend** '91 thru '95

AMC
- **Jeep CJ** - see JEEP (50020)
- 14020 **Mid-size models** '70 thru '83
- 14025 **(Renault) Alliance & Encore** '83 thru '87

AUDI
- 15020 **4000** all models '80 thru '87
- 15025 **5000** all models '77 thru '83
- 15026 **5000** all models '84 thru '88

AUSTIN-HEALEY
- **Sprite** - see MG Midget (66015)

BMW
- *18020 **3/5 Series** not including diesel or all-wheel drive models '82 thru '92
- 18021 **3-Series** incl. Z3 models '92 thru '98
- 18025 **320i** all 4 cyl models '75 thru '83
- 18050 **1500 thru 2002** except Turbo '59 thru '77

BUICK
- *19010 **Buick Century** '97 thru '02
- **Century (front-wheel drive)** - see GM (38005)
- *19020 **Buick, Oldsmobile & Pontiac Full-size (Front-wheel drive)** '85 thru '02
- **Buick** Electra, LeSabre and Park Avenue; **Oldsmobile** Delta 88 Royale, Ninety Eight and Regency; **Pontiac** Bonneville
- 19025 **Buick Oldsmobile & Pontiac Full-size (Rear wheel drive)**
- **Buick** Estate '70 thru '90, Electra '70 thru '84, LeSabre '70 thru '85, Limited '74 thru '79
- **Oldsmobile** Custom Cruiser '70 thru '90, Delta 88 '70 thru '85, Ninety-eight '70 thru '84
- **Pontiac** Bonneville '70 thru '81, Catalina '70 thru '81, Grandville '70 thru '75, Parisienne '83 thru '86
- 19030 **Mid-size Regal & Century** all rear-drive models with V6, V8 and Turbo '74 thru '87
- **Regal** - see GENERAL MOTORS (38010)
- **Riviera** - see GENERAL MOTORS (38030)
- **Roadmaster** - see CHEVROLET (24046)
- **Skyhawk** - see GENERAL MOTORS (38015)
- **Skylark** - see GM (38020, 38025)
- **Somerset** - see GM (38025)

CADILLAC
- 21030 **Cadillac Rear Wheel Drive** all gasoline models '70 thru '93
- **Cimarron** - see GENERAL MOTORS (38015)
- **DeVille** - see GM (38031 & 38032)
- **Eldorado** - see GM (38030 & 38031)
- **Fleetwood** - see GM (38031)
- **Seville** - see GM (38030, 38031 & 38032)

CHEVROLET
- *24010 **Astro & GMC Safari Mini-vans** '85 thru '02
- 24015 **Camaro V8** all models '70 thru '81
- 24016 **Camaro** all models '82 thru '92
- 24017 **Camaro & Firebird** '93 thru '00
- **Cavalier** - see GENERAL MOTORS (38016)
- **Celebrity** - see GENERAL MOTORS (38005)
- 24020 **Chevelle, Malibu & El Camino** '69 thru '87
- 24024 **Chevette & Pontiac T1000** '76 thru '87
- **Citation** - see GENERAL MOTORS (38020)
- 24032 **Corsica/Beretta** all models '87 thru '96
- 24040 **Corvette** all V8 models '68 thru '82
- 24041 **Corvette** all models '84 thru '96
- 10305 **Chevrolet Engine Overhaul Manual**
- 24045 **Full-size Sedans** Caprice, Impala, Biscayne, Bel Air & Wagons '69 thru '90
- 24046 **Impala SS & Caprice and Buick Roadmaster** '91 thru '96
- **Impala** - see LUMINA (24048)
- **Lumina** '90 thru '94 - see GM (38010)
- *24048 **Lumina & Monte Carlo** '95 thru '01
- **Lumina APV** - see GM (38035)
- 24050 **Luv Pick-up** all 2WD & 4WD '72 thru '82
- **Malibu** '97 thru '00 - see GM (38026)

- 24055 **Monte Carlo** all models '70 thru '88
- **Monte Carlo** '95 thru '01 - see LUMINA (24048)
- 24059 **Nova** all V8 models '69 thru '79
- 24060 **Nova and Geo Prizm** '85 thru '92
- 24064 **Pick-ups '67 thru '87** - Chevrolet & GMC, all V8 & in-line 6 cyl, 2WD & 4WD '67 thru '87; Suburbans, Blazers & Jimmys '67 thru '91
- *24065 **Pick-ups '88 thru '98** - Chevrolet & GMC, full-size pick-ups '88 thru '98, C/K Classic '99 & '00, Blazer & Jimmy '92 thru '94; Suburban '92 thru '99; Tahoe & Yukon '95 thru '99
- *24066 **Pick-ups '99 thru '01** - Chevrolet Silverado & GMC Sierra full-size pick-ups '99 thru '01, Suburban/Tahoe/Yukon/Yukon XL '00 thru '01
- 24070 **S-10 & S-15 Pick-ups** '82 thru '93, **Blazer & Jimmy** '83 thru '94,
- *24071 **S-10 & S-15 Pick-ups** '94 thru '01, **Blazer & Jimmy** '95 thru '01, **Hombre** '96 thru '01
- *24072 **Trailblazer & GMC Envoy** '02 thru '03
- 24075 **Sprint** '85 thru '88 & **Geo Metro** '89 thru '01
- 24080 **Vans** - Chevrolet & GMC '68 thru '96

CHRYSLER
- 25015 **Chrysler Cirrus, Dodge Stratus, Plymouth Breeze** '95 thru '00
- 10310 **Chrysler Engine Overhaul Manual**
- 25020 **Full-size Front-Wheel Drive** '88 thru '93
- **K-Cars** - see DODGE Aries (30008)
- **Laser** - see DODGE Daytona (30030)
- 25025 **Chrysler LHS, Concorde, New Yorker, Dodge** Intrepid, **Eagle** Vision, '93 thru '97
- *25026 **Chrysler LHS, Concorde, 300M, Dodge** Intrepid, '98 thru '03
- 25030 **Chrysler & Plymouth Mid-size** front wheel drive '82 thru '95
- **Rear-wheel Drive** - see Dodge (30050)
- *25035 **PT Cruiser** all models '01 thru '03
- *25040 **Chrysler** Sebring, **Dodge** Stratus & Avenger '95 thru '02

DATSUN
- 28005 **200SX** all models '80 thru '83
- 28007 **B-210** all models '73 thru '78
- 28009 **210** all models '79 thru '82
- 28012 **240Z, 260Z & 280Z** Coupe '70 thru '78
- 28014 **280ZX** Coupe & 2+2 '79 thru '83
- **300ZX** - see NISSAN (72010)
- 28016 **310** all models '78 thru '82
- 28018 **510 & PL521 Pick-up** '68 thru '73
- 28020 **510** all models '78 thru '81
- 28022 **620 Series Pick-up** all models '73 thru '79
- **720 Series Pick-up** - see NISSAN (72030)
- 28025 **810/Maxima** all gas models, '77 thru '84

DODGE
- **400 & 600** - see CHRYSLER (25030)
- 30008 **Aries & Plymouth Reliant** '81 thru '89
- 30010 **Caravan & Plymouth Voyager** '84 thru '95
- *30011 **Caravan & Plymouth Voyager** '96 thru '02
- 30012 **Challenger/Plymouth Saporro** '78 thru '83
- 30016 **Colt & Plymouth Champ** '78 thru '87
- 30020 **Dakota Pick-ups** all models '87 thru '96
- *30021 **Durango** '98 & '99, **Dakota** '97 thru '99
- 30025 **Dart, Demon, Plymouth Barracuda, Duster & Valiant** 6 cyl models '67 thru '76
- 30030 **Daytona & Chrysler Laser** '84 thru '89
- **Intrepid** - see CHRYSLER (25025, 25026)
- 30034 **Neon** all models '95 thru '99
- 30036 **Neon** all models '00 thru '03
- 30035 **Omni & Plymouth Horizon** '78 thru '90
- 30040 **Pick-ups** all full-size models '74 thru '93
- *30041 **Pick-ups** all full-size models '94 thru '01
- 30045 **Ram 50/D50 Pick-ups & Raider and Plymouth Arrow Pick-ups** '79 thru '93
- 30050 **Dodge/Plymouth/Chrysler** RWD '71 thru '89
- 30055 **Shadow & Plymouth Sundance** '87 thru '94
- 30060 **Spirit & Plymouth Acclaim** '89 thru '95
- *30065 **Vans - Dodge & Plymouth** '71 thru '99

EAGLE
- **Talon** - see MITSUBISHI (68030, 68031)
- **Vision** - see CHRYSLER (25025)

FIAT
- 34010 **124 Sport Coupe & Spider** '68 thru '78
- 34025 **X1/9** all models '74 thru '80

FORD
- 10355 **Ford Automatic Transmission Overhaul**
- 36004 **Aerostar Mini-vans** all models '86 thru '97
- 36006 **Contour & Mercury Mystique** '95 thru '00
- 36008 **Courier Pick-up** all models '72 thru '82
- *36012 **Crown Victoria & Mercury Grand Marquis** '88 thru '00
- 10320 **Ford Engine Overhaul Manual**
- 36016 **Escort/Mercury Lynx** all models '81 thru '90
- 36020 **Escort/Mercury Tracer** '91 thru '00
- 36024 **Explorer & Mazda Navajo** '91 thru '01
- 36028 **Fairmont & Mercury Zephyr** '78 thru '83
- 36030 **Festiva & Aspire** '88 thru '97
- 36032 **Fiesta** all models '77 thru '80
- *36034 **Focus** all models '00 and '01
- 36036 **Ford & Mercury Full-size** '75 thru '87
- 36040 **Granada & Mercury Monarch** '75 thru '80
- 36044 **Ford & Mercury Mid-size** '75 thru '86
- 36048 **Mustang V8** all models '64-1/2 thru '73
- 36049 **Mustang II** 4 cyl, V6 & V8 models '74 thru '78
- 36050 **Mustang & Mercury Capri** all models Mustang, '79 thru '93; Capri, '79 thru '86
- *36051 **Mustang** all models '94 thru '00
- 36054 **Pick-ups & Bronco** '73 thru '79
- 36058 **Pick-ups & Bronco** '80 thru '96
- *36059 **F-150 & Expedition** '97 thru '02, **F-250** '97 thru '99 & **Lincoln Navigator** '98 thru '02
- *36060 **Super Duty Pick-ups, Excursion** '97 thru '02
- 36062 **Pinto & Mercury Bobcat** '75 thru '80
- 36066 **Probe** all models '89 thru '92
- 36070 **Ranger/Bronco II** gas models '83 thru '92
- *36071 **Ranger** '93 thru '00 & **Mazda Pick-ups** '94 thru '00
- 36074 **Taurus & Mercury Sable** '86 thru '95
- *36075 **Taurus & Mercury Sable** '96 thru '01
- 36078 **Tempo & Mercury Topaz** '84 thru '94
- 36082 **Thunderbird/Mercury Cougar** '83 thru '88
- 36086 **Thunderbird/Mercury Cougar** '89 and '97
- 36090 **Vans** all V8 Econoline models '69 thru '91
- *36094 **Vans** full size '92 thru '01
- *36097 **Windstar Mini-van** '95 thru '01

GENERAL MOTORS
- 10360 **GM Automatic Transmission Overhaul**
- 38005 **Buick Century, Chevrolet Celebrity, Oldsmobile Cutlass Ciera & Pontiac 6000** all models '82 thru '96
- *38010 **Buick Regal, Chevrolet Lumina, Oldsmobile Cutlass Supreme & Pontiac Grand Prix (FWD)** '88 thru '02
- 38015 **Buick Skyhawk, Cadillac Cimarron, Chevrolet Cavalier, Oldsmobile Firenza & Pontiac J-2000 & Sunbird** '82 thru '94
- *38016 **Chevrolet Cavalier & Pontiac Sunfire** '95 thru '00
- 38020 **Buick Skylark, Chevrolet Citation, Olds Omega, Pontiac Phoenix** '80 thru '85
- 38025 **Buick Skylark & Somerset, Oldsmobile Achieva & Calais and Pontiac Grand Am** all models '85 thru '98
- *38026 **Chevrolet Malibu, Olds Alero & Cutlass, Pontiac Grand Am** '97 thru '00
- 38030 **Cadillac Eldorado** '71 thru '85, **Seville** '80 thru '85, **Oldsmobile Toronado** '71 thru '85, **Buick Riviera** '79 thru '85
- *38031 **Cadillac Eldorado & Seville** '86 thru '91, **DeVille** '86 thru '93, **Fleetwood & Olds Toronado** '86 thru '92, **Buick Riviera** '86 thru '93
- 38032 **Cadillac DeVille** '94 thru '02 & **Seville** - '92 thru '02
- 38035 **Chevrolet Lumina APV, Olds Silhouette & Pontiac Trans Sport** all models '90 thru '95
- *38036 **Chevrolet Venture, Olds Silhouette, Pontiac Trans Sport & Montana** '97 thru '01
- **General Motors Full-size Rear-wheel Drive** - see BUICK (19025)

Listings shown with an asterisk (*) indicate model coverage as of this printing. These titles will be periodically updated to include later model years - consult your Haynes dealer for more information.

Haynes North America, Inc., 861 Lawrence Drive, Newbury Park, CA 91320-1514 • (805) 498-6703

HAYNES REPAIR MANUALS

GEO
- **Metro** - see CHEVROLET Sprint (24075)
- **Prizm** - '85 thru '92 see CHEVY (24060), '93 thru '02 see TOYOTA Corolla (92036)
- 40030 **Storm** all models '90 thru '93
- **Tracker** - see SUZUKI Samurai (90010)

GMC
- **Vans & Pick-ups** - see CHEVROLET

HONDA
- 42010 **Accord CVCC** all models '76 thru '83
- 42011 **Accord** all models '84 thru '89
- 42012 **Accord** all models '90 thru '93
- 42013 **Accord** all models '94 thru '97
- *42014 **Accord** all models '98 and '99
- 42020 **Civic 1200** all models '73 thru '79
- 42021 **Civic 1300 & 1500 CVCC** all models '80 thru '83
- 42022 **Civic 1500 CVCC** all models '75 thru '79
- 42023 **Civic** all models '84 thru '91
- 42024 **Civic & del Sol** '92 thru '95
- *42025 **Civic** '96 thru '00, **CR-V** '97 thru '00, **Acura Integra** '94 thru '00
- 42040 **Prelude CVCC** all models '79 thru '89

HYUNDAI
- *43010 **Elantra** all models '96 thru '01
- 43015 **Excel & Accent** all models '86 thru '98

ISUZU
- **Hombre** - see CHEVROLET S-10 (24071)
- *47017 **Rodeo** '91 thru '02; **Amigo** '89 thru '94 and '98 thru '02; **Honda Passport** '95 thru '02
- 47020 **Trooper & Pick-up** '81 thru '93

JAGUAR
- 49010 **XJ6** all 6 cyl models '68 thru '86
- 49011 **XJ6** all models '88 thru '94
- 49015 **XJ12 & XJS** all 12 cyl models '72 thru '85

JEEP
- 50010 **Cherokee, Comanche & Wagoneer Limited** all models '84 thru '00
- 50020 **CJ** all models '49 thru '86
- *50025 **Grand Cherokee** all models '93 thru '00
- 50029 **Grand Wagoneer & Pick-up** '72 thru '91 Grand Wagoneer '84 thru '91, Cherokee & Wagoneer '72 thru '83, Pick-up '72 thru '88
- *50030 **Wrangler** all models '87 thru '00

LEXUS
- **ES 300** - see TOYOTA Camry (92007)

LINCOLN
- **Navigator** - see FORD Pick-up (36059)
- *59010 **Rear-Wheel Drive** all models '70 thru '01

MAZDA
- 61010 **GLC Hatchback (rear-wheel drive)** '77 thru '83
- 61011 **GLC (front-wheel drive)** '81 thru '85
- 61015 **323 & Protegé** '90 thru '00
- *61016 **MX-5 Miata** '90 thru '97
- 61020 **MPV** all models '89 thru '94
- **Navajo** - see Ford Explorer (36024)
- 61030 **Pick-ups** '72 thru '93
- **Pick-ups** '94 thru '00 - see Ford Ranger (36071)
- 61035 **RX-7** all models '79 thru '85
- 61036 **RX-7** all models '86 thru '91
- 61040 **626 (rear-wheel drive)** all models '79 thru '82
- 61041 **626/MX-6 (front-wheel drive)** '83 thru '91
- 61042 **626** '93 thru '01, **MX-6/Ford Probe** '93 thru '97

MERCEDES-BENZ
- 63012 **123 Series Diesel** '76 thru '85
- 63015 **190 Series** four-cyl gas models, '84 thru '88
- 63020 **230/250/280** 6 cyl sohc models '68 thru '72
- 63025 **280 123 Series** gasoline models '77 thru '81
- 63030 **350 & 450** all models '71 thru '80

MERCURY
- 64200 **Villager & Nissan Quest** '93 thru '01
- *All other titles, see FORD Listing.*

MG
- 66010 **MGB** Roadster & GT Coupe '62 thru '80
- 66015 **MG Midget, Austin Healey Sprite** '58 thru '80

MITSUBISHI
- 68020 **Cordia, Tredia, Galant, Precis & Mirage** '83 thru '93
- 68030 **Eclipse, Eagle Talon & Ply. Laser** '90 thru '94
- *68031 **Eclipse** '95 thru '01, **Eagle Talon** '95 thru '98
- *68035 **Galant** '94 thru '02
- 68040 **Pick-up** '83 thru '96 & **Montero** '83 thru '93

NISSAN
- 72010 **300ZX** all models including Turbo '84 thru '89
- 72015 **Altima** all models '93 thru '01
- 72020 **Maxima** all models '85 thru '92
- *72021 **Maxima** all models '93 thru '01
- 72030 **Pick-ups** '80 thru '97 **Pathfinder** '87 thru '95
- *72031 **Frontier Pick-up** '98 thru '01, **Xterra** '00 & '01, **Pathfinder** '96 thru '01
- 72040 **Pulsar** all models '83 thru '86
- **Quest** - see MERCURY Villager (64200)
- 72050 **Sentra** all models '82 thru '94
- 72051 **Sentra & 200SX** all models '95 thru '99
- 72060 **Stanza** all models '82 thru '90

OLDSMOBILE
- 73015 **Cutlass** V6 & V8 gas models '74 thru '88
- *For other OLDSMOBILE titles, see BUICK, CHEVROLET or GENERAL MOTORS listing.*

PLYMOUTH
- *For PLYMOUTH titles, see DODGE listing.*

PONTIAC
- 79008 **Fiero** all models '84 thru '88
- 79018 **Firebird** V8 models except Turbo '70 thru '81
- 79019 **Firebird** all models '82 thru '92
- 79040 **Mid-size Rear-wheel Drive** '70 thru '87
- *For other PONTIAC titles, see BUICK, CHEVROLET or GENERAL MOTORS listing.*

PORSCHE
- 80020 **911** except Turbo & Carrera 4 '65 thru '89
- 80025 **914** all 4 cyl models '69 thru '76
- 80030 **924** all models including Turbo '76 thru '82
- 80035 **944** all models including Turbo '83 thru '89

RENAULT
- **Alliance & Encore** - see AMC (14020)

SAAB
- *84010 **900** all models including Turbo '79 thru '88

SATURN
- *87010 **Saturn** all models '91 thru '99

SUBARU
- 89002 **1100, 1300, 1400 & 1600** '71 thru '79
- 89003 **1600 & 1800** 2WD & 4WD '80 thru '94

SUZUKI
- 90010 **Samurai/Sidekick & Geo Tracker** '86 thru '01

TOYOTA
- 92005 **Camry** all models '83 thru '91
- 92006 **Camry** all models '92 thru '96
- *92007 **Camry, Avalon, Solara, Lexus ES 300** '97 thru '01
- 92015 **Celica Rear Wheel Drive** '71 thru '85
- 92020 **Celica Front Wheel Drive** '86 thru '99
- 92025 **Celica Supra** all models '79 thru '92
- 92030 **Corolla** all models '75 thru '79
- 92032 **Corolla** all rear wheel drive models '80 thru '87
- 92035 **Corolla** all front wheel drive models '84 thru '92
- 92036 **Corolla & Geo Prizm** '93 thru '02
- 92040 **Corolla Tercel** all models '80 thru '82
- 92045 **Corona** all models '74 thru '82
- 92050 **Cressida** all models '78 thru '82
- 92055 **Land Cruiser** FJ40, 43, 45, 55 '68 thru '82
- 92056 **Land Cruiser** FJ60, 62, 80, FZJ80 '80 thru '96
- 92065 **MR2** all models '85 thru '87
- 92070 **Pick-up** all models '69 thru '78
- 92075 **Pick-up** all models '79 thru '95
- *92076 **Tacoma** '95 thru '00, **4Runner** '96 thru '00, & **T100** '93 thru '98
- *92078 **Tundra** '00 thru '02 & **Sequoia** '01 thru '02
- 92080 **Previa** all models '91 thru '95
- *92082 **RAV4** all models '96 thru '02
- 92085 **Tercel** all models '87 thru '94
- 92090 **Sienna Van** all models '98 thru '02

TRIUMPH
- 94007 **Spitfire** all models '62 thru '81
- 94010 **TR7** all models '75 thru '81

VW
- 96008 **Beetle & Karmann Ghia** '54 thru '79
- *96009 **New Beetle** '98 thru '00
- 96016 **Rabbit, Jetta, Scirocco & Pick-up** gas models '74 thru '91 & Convertible '80 thru '92
- 96017 **Golf, GTI & Jetta** '93 thru '98 & **Cabrio** '95 thru '98
- *96018 **Golf, GTI, Jetta & Cabrio** '99 thru '02
- 96020 **Rabbit, Jetta & Pick-up** diesel '77 thru '84
- 96023 **Passat** '98 thru '01, **Audi A4** '96 thru '01
- 96030 **Transporter 1600** all models '68 thru '79
- 96035 **Transporter 1700, 1800 & 2000** '72 thru '79
- 96040 **Type 3 1500 & 1600** all models '63 thru '73
- 96045 **Vanagon** all air-cooled models '80 thru '83

VOLVO
- 97010 **120, 130 Series & 1800 Sports** '61 thru '73
- 97015 **140 Series** all models '66 thru '74
- 97020 **240 Series** all models '76 thru '93
- 97040 **740 & 760 Series** all models '82 thru '88
- 97050 **850 Series** all models '93 thru '97

TECHBOOK MANUALS
- 10205 **Automotive Computer Codes**
- 10210 **Automotive Emissions Control Manual**
- 10215 **Fuel Injection Manual, 1978 thru 1985**
- 10220 **Fuel Injection Manual, 1986 thru 1999**
- 10225 **Holley Carburetor Manual**
- 10230 **Rochester Carburetor Manual**
- 10240 **Weber/Zenith/Stromberg/SU Carburetors**
- 10305 **Chevrolet Engine Overhaul Manual**
- 10310 **Chrysler Engine Overhaul Manual**
- 10320 **Ford Engine Overhaul Manual**
- 10330 **GM & Ford Diesel Engine Repair Manual**
- 10340 **Small Engine Repair Manual,** 5 HP & Less
- 10341 **Small Engine Repair Manual,** 5.5 - 20 HP
- 10345 **Suspension, Steering & Driveline Manual**
- 10355 **Ford Automatic Transmission Overhaul**
- 10360 **GM Automatic Transmission Overhaul**
- 10405 **Automotive Body Repair & Painting**
- 10410 **Automotive Brake Manual**
- 10411 **Automotive Anti-lock Brake (ABS) Systems**
- 10415 **Automotive Detaiing Manual**
- 10420 **Automotive Eelectrical Manual**
- 10425 **Automotive Heating & Air Conditioning**
- 10430 **Automotive Reference Manual & Dictionary**
- 10435 **Automotive Tools Manual**
- 10440 **Used Car Buying Guide**
- 10445 **Welding Manual**
- 10450 **ATV Basics**

SPANISH MANUALS
- 98903 **Reparación de Carrocería & Pintura**
- 98905 **Códigos Automotrices de la Computadora**
- 98910 **Frenos Automotriz**
- 98915 **Inyección de Combustible 1986 al 1999**
- 99040 **Chevrolet & GMC Camionetas** '67 al '87 Incluye Suburban, Blazer & Jimmy '67 al '91
- 99041 **Chevrolet & GMC Camionetas** '88 al '98 Incluye Suburban '92 al '98, Blazer & Jimmy '92 al '94, Tahoe y Yukon '95 al '98
- 99042 **Chevrolet & GMC Camionetas Cerradas** '68 al '95
- 99055 **Dodge Caravan & Plymouth Voyager** '84 al '95
- 99075 **Ford Camionetas y Bronco** '80 al '94
- 99077 **Ford Camionetas Cerradas** '69 al '91
- 99083 **Ford Modelos de Tamaño Grande** '75 al '87
- 99088 **Ford Modelos de Tamaño Mediano** '75 al '86
- 99091 **Ford Taurus & Mercury Sable** '86 al '95
- 99095 **GM Modelos de Tamaño Grande** '70 al '90
- 99100 **GM Modelos de Tamaño Mediano** '70 al '88
- 99110 **Nissan Camioneta** '80 al '96, **Pathfinder** '87 al '95
- 99118 **Nissan Sentra** '82 al '94
- 99125 **Toyota Camionetas y 4Runner** '79 al '95

Listings shown with an asterisk (*) indicate model coverage as of this printing. These titles will be periodically updated to include later model years - consult your Haynes dealer for more information.

Haynes North America, Inc., 861 Lawrence Drive, Newbury Park, CA 91320-1514 • (805) 498-6703